YOUR FIRST

OWNING, DRIVING, MAINTAINING

ROLAND JONES

First published 1993

ISBN 0 7110 2092 2

All rights reserved. No part of this book may be reproduced or transmitted in any form or by any means, electronic or mechanical, including photocopying, recording or by any information storage and retrieval system, without permission from the Publisher in writing.

© R. T. Jones 1993

Published by Ian Allan Ltd, Shepperton, Surrey; and printed by Ian Allan Printing Ltd at their works at Coombelands in Runnymede, England.

Cover drawing by Robert Heesom

Acknowledgements

The Author wishes to thank the following for their help and co-operation in compiling this book and for permission to reproduce a number of their photographs:

The Automobile Association (AA), The Royal Automobile Club (RAC), Rolls Royce Motor Cars Ltd., Rover Lex (Bexleyheath), Ford Motor Co, Ltd., Star Motors (Welling), Laidlaws of Bexley, Kentish Times Newspapers, Peter Hull – Vintage Sports Club Ltd., Lotus Cars (Norwich), Mercedes Benz, London Fire and Civil Defence Authority (LFCDA), Auto Repairs (Bexley Village), Rover Cars, ADT Auctions, Volkswagen (GB) Ltd., Wyndham Leigh Ltd., Snowchains Europroducts, Technorizon International Plc, Carchair Ltd., MLP Insurance (St Albans).

My family and friends, Anne, Trevor, David, Colin, Lucinda and Linda Pearson, Darren McKay, Brian Gibbs and Dave Lane.

Her Majesty's Stationery Office for permission to reproduce facsimiles of Department of Transport Forms.

For the purpose of the book some photographic details have had to be simulated, controlled or pre-arranged i.e. breakdowns, car park sequences etc, for which I am indebted to my friends.

Foreword

The aim of this book is to assist people who are learning to drive, or those who have recently passed their driving test and are new to motoring. It outlines what car owners need to do in order to comply with current legislation and what maintenance is required in order to obtain the best performance and reliability from a car.

Changes are constantly being made to vehicles and their operating conditions and many changes in legislation are envisaged which all takes time. However, at the time of writing the Author has done his best to be up-to-date with his information.

Contents

289 WNO

1 Buying a Car

Below:

'I've done it!' All drivers are familiar with this feeling of elation. However, new drivers should proceed with caution and only after gaining experience on the road should they regard themselves as a competent and accomplished driver .

INTRODUCTION

Before buying a car it is worth remembering the following points:
1. The initial purchase price is not necessarily the most important factor.
2. Running costs and maintenance are an important consideration.
3. Road Tax has to be paid and this is the same for an old 'banger' as it is for a new car.
4. Insurance is particularly expensive for young first time drivers and the make of car can affect premiums drastically.
5. Money should be kept in reserve for items such as batteries, tyres, and exhaust systems which, sooner or later, will need to be replaced.

Some petrol companies are now introducing petrol which they claim cleans the engine and is 13% cleaner and kinder to the atmosphere.

Legislation is in hand to ensure that by about 1993 all new cars will have catalytic converters fitted to their exhaust systems.

These will cut down the engine's output of harmful gases which are the by-products of combustion. They include carbon monoxide — which is poisonous — hydrocarbons, which are the smog producers, and the oxides of nitrogens, which are the acid-rain makers.

For those concerned about the environment, another consideration may be whether or not the car can run on unleaded petrol, which is now widely available.

Unleaded petrol, however, should only be used on cars tuned for its use and those fitted with a catalytic converter. Leaded petrols will ruin a converter.

CHOOSING A CAR

If you can afford a new car, you will be spared the problems usually associated with buying a secondhand vehicle and can concentrate on choosing the right model for your requirements. Try and ensure that you buy from a franchised dealer and not from a manufacturer

Right:

A model such as this with two side doors and a tail-gate is referred to as a three-door hatchback. They give good petrol consumption and are relatively cheaper to run.

specialist. A franchised dealer has an agreement with a manufacturer to sell, service and repair cars in its product range, but a specialist is simply a garage that deals mainly in one make and generally has no link with the manufacturer.

Work done by a specialist or non-franchised dealer, even if well executed, can invalidate a warranty which requires that a car is serviced by a franchised dealer only.

Your Requirements

Before deciding upon a model you will need to consider your particular requirements. For example, how many passengers do you usually carry at any one time? Do you often give lifts to friends who may be accompanied by their children, or do you take elderly parents regularly to the local hospital for example and will they find the model you choose easy to get in and out of? Do you have a dog with you most of the time, so do you need a hatchback, or would an estate model be more suitable? Do you ever tow anything — having a tow-bar fitted can be quite expensive and some models are not easily adapted.

Do you need a car with a decent weight-carrying capacity? Are most of your journeys likely to be through town traffic, perhaps to your work place?

Left:
This estate car is particularly suitable for carrying equipment and has ample room for conveying pets which can be separated — by a grill — from passengers.

What Transmission

Do you prefer to use a manually operated gearbox or an automatic transmission gearbox? Many new drivers often keep to the type of car on which they learned as they feel more comfortable with it. Of course, those who learned on an automatic will have no choice as their licence only enables them to drive vehicles with automatic transmission.

What Engine

There are two main types of engine — petrol and diesel, with variations. Very briefly, petrol-powered cars are cheaper to buy, dearer to run, smoother running, have faster acceleration and are quieter.

Diesel-powered cars on the other hand are dearer to buy, cheaper to run as the engine is at least 25% more efficient than its petrol-driven counterpart, noisy when idling and appear sluggish when accelerating compared to an equivalent petrol-powered car. They are very reliable and not affected by dampness — so often the cause of petrol engines' unwillingness to start in winter.

Left:
One of the vast numbers of present-day carburetted petrol engines. However, times are changing and they are likely to be superseded by the petrol-injected type.

Right:

The early diesels had a poor reputation and held little appeal for the public but great improvements have been made to the modern engine and it is now seen in a much more favourable light and being bought in ever increasing numbers.

Below:

This petrol-injected engine is proving to be more efficient and an easier starter than the carburetted version. It has an expensive, sophisticated computerised fuel distribution system which gives better fuel consumption and produces less pollutants.

Where you Live

You should consider too the area in which you live. Are there narrow lanes and do mud and leaves lie around for long stretches of the year? Might you need all-year-round tyres which are designed for both summer and winter use and have been particularly developed to combat slippery surfaces?

For those who live in hilly or mountainous regions prolonged, harsh wintry conditions may pose problems. Perhaps a four-wheel-drive model, of which there are an increasing number being manufactured, should be considered. This type of vehicle is a boon in icy conditions, on un-made-up roads or for cross-country work.

Making a Decision

Before making a decision all these requirements should be carefully considered. There is a wide choice of new cars available and it pays to browse through the manufacturers' brochures.

Follow this up with visits to car showrooms where you can ask a salesperson pertinent questions about the model in which you are interested, and do take the opportunity to have a test drive. Take time to arrive at a decision; after all, buying a car these days is an expensive undertaking. Remember also that the depreciation of new cars can be quite substantial after only a year or two. Just driving out of a showroom can reduce the value of the car by £2,000.

Paintwork

When buying a new car you shouldn't need to worry about wear in the new components, but nevertheless complaints are made about new cars and by far the most common faults concern paintwork and rust. Unfortunately, new cars are often parked for months and during this period blisters may develop under the paint and rust start to form in crevices.

When in the car showroom examine the bodywork carefully, looking particularly for rust. Position yourself so that the light falls on the painted surfaces you are inspecting — this will often highlight a paint blemish. Ask the salesman if the garage will guarantee to rectify any blemishes or faults found; if not, reject the car and ask to see another.

Warranty

Because of the depressed state of the motor industry at present manufacturers are offering very competitive warranties with their cars. They are able to do this partly because of increased confidence due to improved design and better corrosion-proofing measures.

A typical warranty might include complete servicing and a full rectification on any faults developed in twelve months of unlimited mileage; a six year anti-corrosion warranty; a three-year cosmetic paint warranty on certain products; AA or RAC membership benefits– and half-price rates for other members of your household.

After this first year, there is an option to take out, at a modest cost, an additional warranty for the second and third year. The majority of the car's components are insured against most mechanical and electrical problems, including the cost of labour. Your AA or RAC membership benefits also continue, and you are given car-hire contributions if your car is off the road for more than 24 hours. There may also be other benefits that can be negotiated with the deal.

Finance

If you are not paying for the car by cash or cheque there are banks, finance houses and perhaps other sources from which you can raise a loan.

Above:

A visit to a car showroom with its array of gleaming new models can be quite exciting, especially if you know that one of them may shortly be yours. Nevertheless, look closely for signs of rust or paint blemishes.

Left:

This potential buyer, having looked at the exterior for any obvious dents or rust, is now checking the condition of the interior. Inspecting the state of the seats and condition of the headlining will give a good indication of whether the vehicle has been carefully used.

Above:
Look closely at the bodywork for any evidence of re-spray work. If a car has been re-sprayed there is usually some tell-tale mark. Ask yourself why a fairly new car should require it? Viewing in natural light is best as artificial light can disguise blemishes. Looking along the body to catch the light pinpoints any faults.

No doubt the dealer from whom you are buying the car will suggest that his organisation can offer an attractive deal. For example, if you are buying a car from Rover they will offer you their Rover Finance Car loan scheme, which consists of a personal loan for a new car or for a car up to five years old. The scheme may also include the road fund licence and vehicle insurance.

Their terms at present, with inflation running at 4.7% are as follows:

For a Metro model:
For 50% deposit 12 months @ 0% APR (Annual Percentage Rate)
For 50% deposit 24 months @ 7.9% APR
For 50% deposit 36 months @ 15.6% APR
For 20% deposit 48 months @ 15.3% APR

Size of Car.

Consider whether you really need a large powerful car, or would it be more prudent, at this stage, to select one of the many smaller models on the market which are more economical. Remember, running a larger car is more expensive: insurance is higher, tyres and batteries are larger and therefore more expensive, and petrol consumption is higher. When you have decided which car suits your needs best, try to ensure value for money.

Nearly New Cars

Money is usually the over-riding factor when buying a car.

If you are able to afford a fairly new second-hand one that has not had its first MOT inspection, (ie under three years old) you will find that most reputable motor agents have a selection of good-quality cars to choose from. These cars have usually been traded-in and you should be looking for a fairly low mileage and excellent condition throughout.

Remember that all organisations will present their cars in the best possible light — the interiors will have been thoroughly cleaned and the bodywork will be gleaming to give a good first impression. Fair enough, but make sure you know its past history and are aware of its mechanical condition.

Inspect the bodywork to see that there are no obvious dents, signs of rust, faults, etc. Look closely for any evidence of re-spray work; a fairly new car should not have been re-sprayed, unless the vehicle had a minor accident and was competently repaired. Although this information should be volunteered by the car salesman, it is as well to draw attention to such matters yourself.

The business of selling cars is so competitive these days that in order to secure a sale the seller is usually willing to offer a discount. After you have spent some time viewing the car you have decided on, press the salesman for a discount and ask what sort of after-sales-service he can offer. If you are buying from a reputable dealer you should get a warranty commensurate with the price you are paying. It is up to you to secure the best deal possible.

Older Cars

If you want a car lower down the price range, greater care is needed in your selection. However, a cheaper car does have its advantages. For instance, if you are still somewhat of a novice driver sustaining a dent, it isn't quite so disastrous as it would be if you were driving a more expensive car. Also, ownership of a cheap, low-insurance category car for a year or two helps build up a No-Claims Discount (See Chapter 3).

OTHER ASPECTS TO CONSIDER

Try and see any car in which you are interested in daylight, as artificial light can disguise blemishes and other defects.

The first things to note when viewing a car is whether the general outside appearance is pleasing and whether there are any obvious defects with the bodywork. An interior inspection should reveal whether or not a vehicle has been used carefully. You would not, for example, expect to see any tears or marks in the roof lining or upholstery in a car up to four or five years old.

Left:

It is a good idea to have someone with you, especially if they have quite a bit of motoring experience behind them.

Far left:

The condition of the pedal rubbers may point to the true mileage.

Left:

Look carefully under the wheel arches – these areas are vulnerable to road grit, salt, ice etc.

Although due allowance should be made for the age of a car, particularly regarding the condition of the driver's seat, carpets, etc, watch for very worn pedal rubbers which may contradict the mileage that the seller claims the vehicle has covered — brand new pedal rubbers are equally suspicious.

Look under the wheel arches, front wings and surrounding body-work as these areas take a lot of punishment from rain, mud, ice, snow and salt thrown up by the wheels. During the winter these areas may stay wet for months, not thoroughly drying until the summer.

Try bearing down rhythmically on all four corners of the car to bounce it up and down. If it bounces more than twice after you have let go, the suspension is worn and may be expensive to rectify.

Mileage

When looking at a car it is important to consider the mileage that has been covered, and it is even more important to know *how* those miles have been covered. As a rule, the ordinary private motorist covers an average of 10,000 miles a year in his car, but 10,000 miles made up of short journeys is a very different matter from the same distance being made over long journeys.

When buying a car always ask about its history. Most private owners will normally offer this information without being asked, but when buying from a dealer it is particularly important to find out about the car's previous usage. Was it a fleet-car? Was it obtained from a private owner? How many owners has it had altogether? However, a secondhand car of higher than average mileage need not necessarily be disregarded. A car which has been used for business purposes, covering a large territory and doing most of its journeying on motorways, may have been less stressed than a car used for constant town journeying. In other words, 1,000 miles of motorway driving can cause less wear than 50 miles of town driving.

This is because a car travelling on a motorway is not continually being stopped and started, needs very few gear or clutch operations and requires few braking efforts, all of which means components such as the brakes, the gears and the clutch are subject to less wear and tear.

Where and How A Car is Used

Next to consider is the area in which the vehicle has been in use. If it has been driven in particularly hilly or even mountainous areas for example then its mileage will have greater significance because the engine will have been used to its limits.

Be especially wary if the car has been used in a coastal area as the environment is particularly unkind to metals, and body-work and

Far left:

Bear down on the suspension to see, upon release, how rapidly it ceases to bounce. A good suspension will not bounce more than twice.

Left:

The mileage of a car that has been used in a region such as this is significant because it will have had relatively more wear and tear, especially in the winter months. A coastal environment is particularly unkind to paint and bodywork. It is important to know whether a car has been garaged or not.

THE SECURITY CODED
RADIO/CASSETTE
TTED TO THIS VEHICLE
LL NOT OPERATE
IF REMOVED

Far left:

Check that the switches and controls operate properly and that they are within comfortable reach for you.

Left:

This prospective buyer is viewing at the seller's address and has obviously been studying prices.

components will need to be carefully inspected for rust and corrosion from salt.

Other Points

Other areas of the car to inspect are doors, locks, windows and lights which should all operate correctly and easily. Tyres, batteries and exhaust systems should be in good condition and the mechanical condition of the car completely sound. Sit in the car and try it for leg and headroom, particularly if you are tall. Check also to see that all switches and controls are within easy reach.

Make sure, too, that there is enough room in the back as some cars have poor leg or head room.

Buying Privately

Buying privately can be a bit of a minefield because private sellers are not governed by the 1979 'Sale of Goods' act which requires goods to be of merchantable quality. It is wise to assess the character of the individual you are dealing with as carefully as the condition of the car, and rely on your instincts.

As has been said before, have a good look round the vehicle, making sensible allowances for imperfections due to genuine wear and tear. Ask to have the engine started up and note whether the exhaust is reasonably clear, or whether it is blue. Allow the engine to idle for a minute or so and then blip the throttle — if a cloud of blue smoke issues from the exhaust pipe this will indicate that the engine is worn and is burning oil. When an engine gets to this stage of wear it needs quite a bit of repair work to get it back to a reasonable condition.

Have a look underneath to see whether the gearbox, back axle, driveshafts and their gaiters, differential or the engine sump are leaking oil. Any leaks from these main components could be quite expensive to put right. If you see any signs of bodyfiller or fibreglass repairs to any structural or load-bearing section, reject the car immediately.

Right:
Check the engine's condition by allowing the engine to idle for a minute or two, then blip the throttle and see whether a cloud of exhaust gas is emitted. This engine is in a poor condition.

Right:
Look under the car to see whether any of the main components are leaking oil.

Price

Where should one look for prices? *Exchange & Mart* is a good starting point as cars are listed alphabetically by make and model; another idea is to look in the local papers. See how much people are asking for the same model and year of registration as the one you are considering, but bear in mind that this can only serve as a rough guide as so much depends on the condition of the car. Alternatively, guides to the selling price of secondhand cars can be purchased from bookstalls and there are many magazines which specialise in publishing current car prices.

When negotiating the purchase price it is in your interest to point out any defects or adverse points concerning the car in order to get a reduction, but expect this to be countered by the seller who will naturally emphasise its plus points and performance.

Always take the opportunity to drive the car so that you can see for yourself how well you manage to handle it before buying, and taking someone else along for a second opinion is recommended.

Documents

Any car over three years of age should have a current MOT (Ministry of Transport — now called the Department of Transport) certificate (see chapter 10). This certificate, among other items, will state the vehicle's mileage as shown on the odometer at the time of the MOT inspection. This will help to corroborate what the seller claims is the vehicle's true mileage.

The Vehicle Registration Document (Log Book) should be given to you when you buy the vehicle (see chapter 2). Check that the

Left:

The buyer is checking the Log Book (form V5) details with the vehicle identification plate.

engine and chassis numbers agree with those shown on the Vehicle Registration Document.

Remember, this document does not prove legal ownership.

Before buying a vehicle you should satisfy yourself that the seller either owns the vehicle, or is entitled to offer it for sale. Ask to see a bill of sale in their name, or other evidence such as a hire-purchase discharge document. It is a good idea to view the vehicle at the address shown on the Registration Document. If the vehicle is not

Far left:

Having made due allowance for the time elapsed since the MOT date, this buyer is checking the MOT certificated mileage with the odometer.

Centre left:

Use a magnet to check areas which may have been repaired or disguised by the use of glass fibre.

Left:

Look closely to see whether an extensive use of underseal is disguising any bad areas.

registered in the seller's name, or a Registration Document is not available, ask why, and if you are in any doubt do not proceed with the sale.

Old Bangers

Many new young drivers turn their attention to the bottom end of the market — to the 'old Banger'. These are often cars which dealers have taken in part-exchange for newer models.

Great care should be exercised with those cars, because they may have been spruced-up and possibly given only the minimum of attention. Bear in mind that all the car components are likely to be worn and some that may need replacing may no longer be in production.

Many body areas are likely to have been repaired in the past with bodyfiller or fibreglass and this can be confirmed by the use of a magnet. Be suspicious of any thickly or recently applied underseal.

Right:
Before an auction commences the public view some of the cars being offered for auction under cover.

AUCTIONS

Auctions are an alternative way of obtaining a car and as the vehicles range enormously you often get a pretty good choice. Whether you are buying or selling, auctions can offer a lot to the private individual and there are different types to choose from. The better-run ones provide good parking and viewing areas and everything is conducted under cover. Refreshments are usually available, and some have restaurants serving full meals. The auctioneer will be a member of the Society of Motor Auctions.

At some smaller auctions the auctioneer may not be a member of the Society, the facilities are often poor and are usually conducted outside. There *are* bargains to be had at these auctions but you need to know what you are doing.

Some of the better-run auctions hold two sessions in a day because of the large numbers of cars to be disposed of. The first session usually consists of cars put in by private sellers and in general the public do the bidding.

The second session, consisting mainly of ex-fleet cars, is usually attended by the Trade.

Attending an auction

As representatives of the motor trade know what they are bidding for and at what price to stop bidding, a lot can be learned from watching them. One of the great advantages of attending an auction is that it saves weeks of driving around to see a number of vehicles. If you are contemplating buying a car in this way you should pay several visits to one to see how they work.

Make notes and list the models that are bid for and how much they fetch. You will then have some idea how much you may have to bid when you next attend. Get there early and seek out four or five likely models which interest you as you are not likely to get the first car that you bid for. Take a look at the information posted on the windscreen, which includes an engineer's inspection if one has been made.

Inspecting the Cars

Inspect the cars as thoroughly as possible.

Wait to hear them started up and make a note of whether they started easily and idled quietly. Look at the exhaust fumes to see whether they are clear. If they are black and smoky, it may mean that the engine requires a decoke or that it is running rich; blue smoke indicates a worn engine. If you have not looked under the bonnet before, now is the chance to ask the driver to release the catch. You cannot really tell much about the condition of an engine just by looking at it, but you can tell whether it has been well maintained or not. If there are any badly corroded battery terminals, or the hoses look in a poor state, or dirt, oil and general filth are present, then this will show that the car has been generally neglected.

Look at the dashboard and instrumentation and watch out for low oil pressure. Ask the driver to blip the accelerator to see whether the oil pressure responds and look at the warning lights and other warning symbols. Note whether the driver is having trouble selecting the gears; is he having to pump the brakes — listen for unusual noises. This is your last chance to evaluate the car's worth to you.

When the bidding starts listen to what the auctioneer has to say about the vehicle and make your bid accordingly. There are bargains to be had if you do your homework beforehand.

In the Ring

When a car enters the ring the auctioneer states its year, make and model. He says whether it is being sold 'as seen' with all its faults or 'on description', also referred to as 'all good' or 'no major mechanical defects'.

'As seen' is generally applied to cars over five years old which realise any amount up to £1,000. These cars are sold with all their

Left:
Many cars are often still within their warranty period.

Below:
Some owners like an engineer's report to be exhibited on their car.

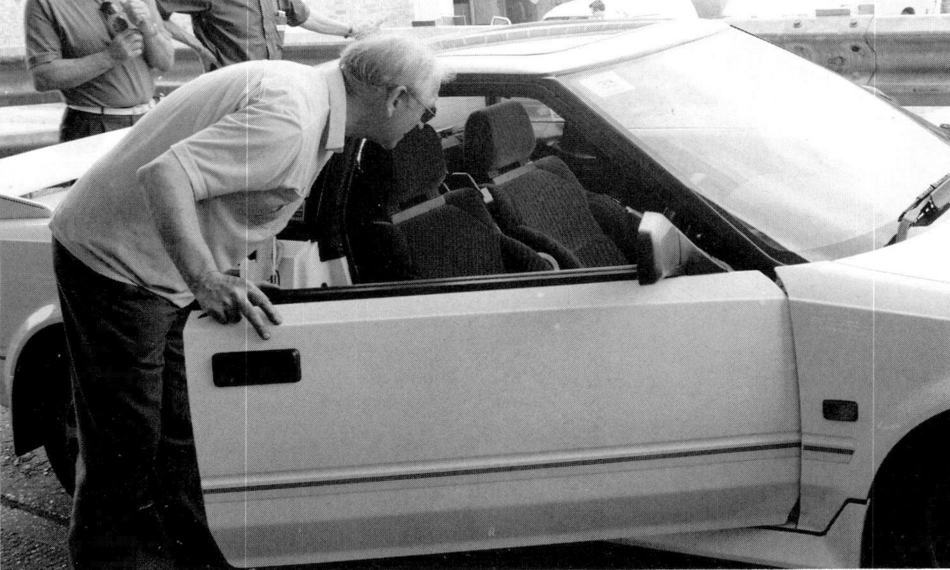

faults, whether major or minor. You must conduct a very careful inspection when buying 'as seen' so that you buy wisely.

A car sold 'all good' is not necessarily faultless, but it should not have any seriously defective main components. You will have one hour after you have bought the car, or after the auction ends to discover any faults; report them immediately and your money will be refunded or the auctioneer will negotiate a lower price with the vendor on your behalf. However, this does not apply if an auctioneer has declared a specific fault, such as 'this one has a noisy back axle' which you have missed because you were not listening attentively enough.

If an 'all good' is being sold with a particular warranted mileage, the auctioneer will declare this as well. If, subsequently, you should find this to be false you are entitled to a refund.

Warranty

At some auctions cars in the 'all good' category are offered with the option of an insurance based extended warranty. This is usually the case at auctions which operate as members of The Society of Motor Auctions. The option may or may not be displayed on the windscreen. You should enquire to see whether one is available. These warranties are very similar to those offered by car dealers and take a lot of the risk out of buying at auction.

Other Auctions

Smaller auctions may only sell perhaps 40 cars a week and most of the cars will have come from private sellers.

Beware of any person who tries to strike up a bargain with you before the car enters the sale, even though he may have a log book and offer a reasonable explanation for his action, such as he doesn't want to pay commission and therefore can let the car go cheaper.

Remember a Log Book is not proof of ownership so be on your guard. Having said this, there are bargains to be had, but you really do need to know what you are doing before venturing to bid at these smaller auctions.

Payment And Indemnity Fee

At the fall of the hammer you will have to pay 10% of the bid price (or £300 if the purchase price is less than £3,000) and you will also be charged an indemnity fee. The indemnity fee at present is £16, payable on the first £1,000 and then £2 per every £1,000 after that.

This is a legal requirement which affords protection to purchasers against the following risks:

1. If the vehicle is later proved to have been stolen property.
2. If it is the subject of a still-outstanding hire purchase agreement or a similar arrangement.

3. If the warranted odometer reading is subsequently proved to be false.
4. If the vehicle has been the subject of an insurance company's total lost settlement, before the date of the auction and this fact has not been disclosed.

In general after payment, the following procedure is adopted by auctioneers up and down the country.

If you are paying by cash or by banker's order you can remove the car straightaway, but it is incumbent on you to tax and insure the car before taking it on the road.

Upon payment you will be given a receipt (for the purchase), the Vehicle Registration Document (Log Book), any MOT certificate if appropriate and a pass-out to take the vehicle off the premises. The documents, when shown to the 'Key' Office, will allow you to obtain the keys to the car. The pass-out, when surrendered at the gate, will enable you to leave.

If you pay by cheque the car remains with the auctioneer until the cheque is cleared, which is usually three working days. You will most likely be charged a parking fee for that time. The above procedure then follows.

Facing page, top:
Taking a last look under the bonnet before the car enters the ring.

Facing page, bottom:
Look at the instrumentation and inspect the general condition of the interior.

Left:
A car in the ring. Bidding commences briskly; you have to be positive to catch the auctioneer's eye, otherwise you may miss your opportunity to bid.

Right:

An AA information and recruiting office, of which there are many situated at prominent sites all over the country.

Far right:

This RAC vehicle inspector is carrying out a comprehensive car inspection for a prospective buyer and will also give an assessment as to its value.

MOTORING ORGANISATIONS

New drivers who are not technically minded may like to consider becoming a member of either The Royal Automobile Club (RAC) or The Automobile Association (AA). Their aim is to make the motorist's life easier and to be readily available 24 hours a day in times of distress or vehicle breakdown.

Both are national organisations which means help should never be too far away, if their own patrol men cannot get to you they will arrange for someone from one of their appointed garages to help you.

Each organisation also offers a complete technical inspection of any vehicle you may be considering purchasing and will give you a comprehensive report and an authoritative opinion as to its value.

The RAC and the AA also offer many other beneficial services to their members, including car insurance, weather information, legal advice and information about road conditions and traffic problems around the country.

There are other organisations which mainly cover breakdowns and get-you-home services, but these are smaller and not able to offer such comprehensive benefits.

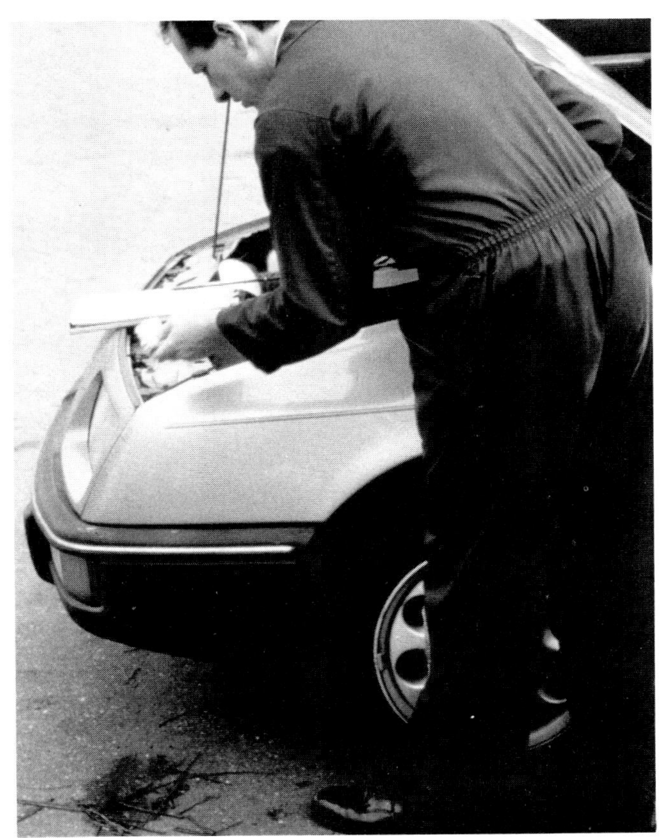

The Future

It seems that the rescue services are considering a quite futuristic idea to be able to pinpoint precisely where a motorist is when his car breaks down.

Very often when a motorist's car does suffer a breakdown he does not always have a precise idea just where he is. This is particularly true when he is undertaking a long journey, perhaps going on holiday. It means at times like this he is not very helpful to the rescue people with his rather vague directions, and a lot of time is lost in trying to locate him.

The new idea will be to place beacons at strategic points all over the country so that should a car break down it will always be within a triangle of beacons. Beams from any three beacons will pick up a directional signal being emitted from the breakdown, relaying its precise position back to HQ; in spite of the fact that the driver may have no idea where he is.

But all of this relies on the manufacturers fitting the appropriate computerised equipment in future to new cars – and at what cost?

DISABLED DRIVERS

For the disabled driver the Department of Transport has set up the Mobility Advice and Vehicle Information Service (MAVIS) to provide advice on driving, car adaptations and car choice both for disabled drivers and for the carriage of disabled passengers.

Its literature gives details on discounts available on new cars purchased by the disabled, Insurance and Excise Tax concessions and other subjects particularly applicable to the disabled.

It also provides an opportunity for disabled persons to try out one or more of a wide range of adapted vehicles at its centre.

The Orange Badge Scheme

The Orange Badge Scheme is a national arrangement which allows disabled people who travel, whether they be driver or passenger, to park as near as possible to their destination. The concessions apply only to on-street parking.

People are entitled to a badge if: they are blind; receive a Mobility Allowance; use a vehicle supplied by the Government; cannot walk without great difficulty; or receive a grant towards running their own vehicle.

A badge is issued to an individual not to a vehicle, so it can be displayed on any vehicle, including taxis and hired vehicles, in which the badge holder is travelling.

The badge must be displayed on the nearside of the windscreen when using the parking benefits and should be removed at all other times.

Badge-holders may park free-of-charge and without time-limit at street parking meters and other time-limit zones.

A special Parking Disc, available from the Authority issuing your badge, may enable you to park on yellow lines for up to two hours, (at present) but you must indicate the time of your arrival on the disc.

You may not park on single or double yellow lines if there is a ban on loading or unloading.

In England and Wales you can apply for a badge to your Social Services Department of your local County District or Borough

Right:
Many firms adapt standard production cars to suit the disabilities of drivers or passengers. The arrangement here shows that this passenger can gain access without having to leave his wheelchair.
Courtesy Carchair Ltd

Above centre:
This car has been specially adapted to suit a driver whose main disability means that she cannot reach the steering wheel. The mechanism here has been designed so that she can operate the steering by rotating the wheel situated between the seats with her left hand.

Below centre:
An orange badge should only be displayed (on nearside of screen) when using street parking benefits. The lower parking disc should show the time parking commenced and should be displayed on the screen near the kerbside. Parking on double yellow lines is permitted but not if there is a ban on loading or unloading.

Far right:
Some of the literature available for the disabled from the Department of Transport's MAVIS Service.

Council. In Scotland you can apply for a badge to the Chief Executive of your local Regional or Island Council. Badges last for three years.

You must not lend your badge to an abled-bodied person for their own use. Misuse of the badge can bring a maximum penalty of a £400 fine.

Further information can be obtained from:
MAVIS — Dept of Transport,
Transport and Road Research Laboratory,
Crowthorne, Berkshire RG11 6AV

Note
Change to the Orange Badge Scheme is being considered by the Department of Transport but the aforesaid is correct at the time of writing.

Proposed changes are that the period for parking on yellow lines be increased from 2-3 hours; and the issuing of a redesigned Orange Badge as a personal passport-type document with space for a photograph of the holder.

INS
and
OUTS
OF CAR CHOICE
Department of Transport
DOOR
to
DOOR
DEPARTMENT
OF TRANSPORT
Mobility
Advice and
Vehicle
Information
Service

2 Vehicle Registration

Department of Transport U 0375066

V5
Rev. Jan/88

Vehicle Registration Document

Registration Mark: B289 WNO
Validation Character: D

JOSEPH BLOGGS
70 THE CRESCENT
TOWN
COUNTY ABC 123

PLEASE QUOTE THE REGISTRATION MARK IN ALL CORRESPONDENCE

Taxation Class	PRIVATE/LIGHT GOODS (PLG)
Make	FORD
Model/Type	ESCORT L ESTATE
Colour(s)	RED
Type of Fuel	HEAVY OIL
VIN/Chassis/Frame No.	WFONXXGCANEL33135
Engine No.	EL33135
Cylinder Capacity	1608 CC
Seating Capacity	
Taxable Weight	
Date of Registration	10 10 84
Last Change of Keeper	13 06 85
No. of Former Keepers	1

This is the Registration Document for the vehicle described opposite. The person named above is the Registered Keeper of the vehicle (the person recorded as keeping it). **The Registered Keeper is not necessarily the legal owner.** This Document is issued by the Driver and Vehicle Licensing Centre on behalf of the Secretary of State for Transport. Police officers and certain officers of the Department of Transport may require you to produce it for inspection at any reasonable time.
YOU ARE REQUIRED BY LAW TO NOTIFY CHANGES TO THE NAME AND ADDRESS OR VEHICLE PARTICULARS PRINTED ON THIS DOCUMENT AS SOON AS THEY OCCUR - SEE OVERLEAF.
For further information about registering, licensing, insuring and testing your vehicle, please ask at a post office or Vehicle Registration Office for leaflet V100.

1. PREVIOUS RECORDED KEEPER: FORD MOTOR CO LTD 011381, EAGLE WAY, BRENTWOOD CM13 3BW.
2. DECLARED NEW AT FIRST REGISTRATION IN GREAT BRITAIN BY MANUFACTURER/SOLE IMPORT CONCESSIONAIRE.
3. DUPLICATE REGISTRATION DOCUMENT.
4. IF YOU SELL THE VEHICLE FILL-IN AND RETURN THE TEAR-OFF SLIP BELOW.
5. IF YOU ARE THE NEW KEEPER AND YOUR NAME IS NOT SHOWN ABOVE TELL US NOW BY FILLING IN THE BACK OF THIS FORM. WE WILL THEN SEND YOU A NEW DOCUMENT IN YOUR NAME.

31

Doc Ref No. 8351 569 1901
16 12 88
1091K028***
871090

B289 WNO D

Notification of Sale or Transfer

If you sell or transfer the vehicle to someone else, COMPLETE AND DETACH this section and send it to DVLC SWANSEA SA99 1AR.
Please do this at once - it is in your own interests to do so.
Give the top part of this document to the new keeper so that he can use it to notify that he has acquired the vehicle.

V5/1
Rev. Jan/88

Registration Mark: B289 WNO | D | 01

Date of Sale or Transfer

Official Use Only

If you have sold or transferred the vehicle to a motor dealer or insurance company, tick here

I sold/transferred this vehicle to the person whose name and address I have written opposite, on the date I have given above. I have also given him/her the top portion of this document.

Signature

NOTE: The new keeper must also notify the change in Section 1 overleaf; otherwise a new Registration Document will not be issued.

Name and Address of new Keeper or dealer acquiring vehicle

Mr 1 | Mrs 2 | Miss 3 | *Please tick box or give other title below*

Other title e.g. Dr., Rev., or company name

Christian or forenames

Surname

Address

Post Town

Postcode

Right:
The front face of the Vehicle Registration Document (log book form V5). All registered Keepers receive one of these for their vehicle from the DVLA.

Centre right:
Application form V100 (available from main post offices) explains how to apply for a Log Book.

Far right:
If you do not receive a Log Book (form V5) with your newly purchased car, you should apply to the DVLA for one, using this form (V62).

INTRODUCTION

Before a vehicle can be used on public roads it must be registered with the Driver's and Vehicle Licensing Agency (DVLA) formerly (DVLC) at Swansea. A Vehicle Registration Document (form V5, still widely known as 'the Log Book') will be issued to the person named as the Registered Keeper of the vehicle. The document must be kept by the Registered Keeper until such times as he sells or disposes of the vehicle.

(Vehicle Registration is explained in form V100 — obtainable from Vehicle Licensing Post Offices).

TO REGISTER A VEHICLE

If you buy a new vehicle the dealer will usually arrange for it to be registered for you, but make sure your correct name, address and post code is entered on the application form. A mistake in this can delay your receipt of the Log Book which will be sent to you by post. When you receive it check to make sure all the particulars are correct. Should there be errors, you will need to send the Log Book back to the DVLA.

If you buy a vehicle from a private owner he must give you the top portion of the Log Book. This document does not prove legal ownership, so before buying a vehicle you should satisfy yourself that the seller either owns the vehicle or is entitled to offer it for sale (see Chapter 1). As soon as you become the new Keeper of the vehicle you should fill in the section 1 at the back of the Log Book and forward it to the DVLA.

If for any reason a Log Book is not available, you can apply for one using Form V62, obtainable from any Vehicle Licensing Post Office. A new Log Book (Vehicle Registration Document) is issued to each new Keeper of a vehicle.

V100
Rev. Jan 90

Registering and Licensing your motor vehicle

...some notes to help you

Before you take a motor vehicle on the public roads it must be registered, currently licensed and covered by a valid test certificate (if it needs one), and you must be licensed to drive and have valid insurance covering your use of the vehicle.

In this leaflet abbreviations are used for the Driver and Vehicle Licensing Agency (DVLA) and Vehicle Registration Offices (VROs).

Data Protection

Information held on the DVLA computer vehicle record is obtained and used for the purpose of registering and licensing vehicles, related law enforcement (including road safety) and assistance to the police in the prevention of crime. Information is also held on a computer register, the purpose of which is to assist in crime prevention and the prosecution of offenders in relation to evasion of Vehicle Excise Duty. The DVLA is so registered under the Data Protection Act 1984–further details are available from the Register of Data Users available in major public libraries throughout the UK.

Where to get further help

Ask at a VRO; or if you wish to enquire about a particular application contact the office to which you applied or the Vehicle Enquiry Unit at DVLA, Swansea SA99 1BL. (Telephone: 0792-72134 between 8.15 a.m. and 4.30 p.m. Monday to Thursday, and 8.15 a.m. to 4.00 p.m. Friday.) In addition, the following leaflets available at VROs and from DVLA may be useful.

- **V355** What you need to know about first registering and licensing motor vehicles which have not been previously registered.
- **V526** Taking a vehicle out of Great Britain.
- **PII** Permanent import of vehicles into Great Britain.
- **D100** What you need to know about driver licensing. (Also available at most post offices.)

REMEMBER

Write clearly on application forms.
Give the vehicle registration mark in any enquiry.
Use your postcode.
Avoid the end of the month rush at Post Offices and VROs by relicensing early.

Issued by:
DVLA
Swansea

Vehicle Registration Document.....application form

Department of Transport

Please read these notes carefully before you fill in this form.

When do I use this form?

You can use this form to apply for a Vehicle Registration Document (V5) either

if you only need a Registration Document
OR
if you also need to apply for a vehicle excise licence (tax disc) and you haven't got a Registration Document for the vehicle.

Remember, you can only apply for a Registration Document if you are the keeper of the vehicle.

What if I also need a tax disc?

If you also need a vehicle excise licence (tax disc) you should apply to a licensing post office using form V10. However, if the vehicle is a Heavy Goods Vehicle, you must apply to a Vehicle Registration Office (VRO) using form V85. The forms V10 and V85 tell you how to apply.

What if I only need a Registration Document?

If you only need a Registration Document, send this form to DVLC Swansea SA99 1AR.

When will I get the Registration Document?

All Registration Documents are issued by the Driver and Vehicle Licensing Centre (DVLC) at Swansea. You should get yours within 4 weeks but please allow a little longer before making enquiries about it.

What if I find the original Registration Document?

If you apply for a Registration Document because you thought you'd lost or misplaced the original one, but you later find it, you must send the original back at once to DVLC, Swansea SA99 1AR. Please enclose a short note saying that you have applied for and received another Registration Document.

What if I need further help?

There is a general leaflet V100 which may help you. You can get a copy from post offices. Or you can telephone the Vehicle Enquiry Unit at DVLC on Swansea (0792) 72134 between 8.15am and 4.30pm Monday to Thursday, and 8.15am to 4pm Friday — or write to us using the postcode SA99 1BL. We will be pleased to help you.

V62

Above:
An example of a cherished registration mark (number plate). The owner has bought a registration mark showing his name, and it also happens to include the number one, which adds to its attractiveness and demand. This type of registration mark can be very costly, for example a motorist recently paid £160,000 for a plate such as this.

Lost Log Book

If you lose or misplace your Log Book you can apply for another using form V62, which is available from main post offices or Vehicle Registration Offices (VRO's). The DVLA will send you a new document after checking your application against the vehicle record. Should you later find the original, send it to the DVLA with a letter explaining what has happened.

Once, you have registered with the DVLA they can not only remind you that the vehicle's licence requires renewal, but if necessary they would be able to tell of any possible safety-related defect found to be developing in your type of vehicle.

Should you change your name upon marriage, or for any other reason, you must inform the DVLA using the back of the Log Book. A new document is sent to you free of charge.

If you are unfortunate enough to have your vehicle stolen you should inform the police at once who will notify the DVLA on your behalf. You do not need to notify the DVLA yourself.

Cherished Registration Marks

If you are about to sell a vehicle which is already licensed and registered in your name with the DVLA and it has a registration mark which you cherish and wish to retain, you may have it transferred to another vehicle.

Form V317 explains what you have to do. It states what the current fee is for this transaction and mentions which further forms are necessary to complete the procedure. Form V317 can be obtained from any VR0.

You can also have a registration mark transferred from an existing vehicle without having to buy and register the vehicle. The owner of the existing vehicle must have the vehicle available to prove to the Department of Transport's inspector that such a vehicle exists and he will also be required to sign a statement on the back of form V317.

Select Registrations

The DVLA has just instituted a new fee-paying Select Registrations scheme whereby new car owners can apply to the DVLA or their agents for a registration mark of a personal nature. The new owner is offered the chance for a fee, to choose his own registration mark. This includes a prefix letter denoting the registration year followed by the choice of any number from 1-20, any number ending in 0 from 30-90, any number ending in 00 from 100-900 or any multiple of the same number such as 22 or 555 (excluding 666).

The number is followed by a combination of three letters of your own personal choice (excluding I,Q,Z) subject to them being available.

The scheme was initially introduced to the public on 1st October 1990 and was offered on a first-come-first-served basis. It has since been amended and enlarged to give over half a million unique numbers available in two price bands, A and B, with two tiers in each.

Band A represents those three-letter combinations which spell popular names and, as you would expect, is the costliest band. Price band B represents all other letter combinations available.

A pamphlet explaining Select Registrations and all the necessary fees can be obtained from the DVLA or their agents.

Department of Transport

Please do not write above this line

1

V317
Rev Aug 89

TRANSFER OF REGISTRATION MARK

Application Form

Please read the leaflet then fill in the form in **black ink** and **CAPITAL letters**

DETAILS OF DONOR VEHICLE

(the vehicle which now carries the mark to be transferred)

Vehicle Registration Mark ______ 2

Make ______ Model ______

VIN/Chassis number ______

If you are going to post this form, give the expiry date and period of the licence disc if any ______ 19 ___ ; ______ months

Name and address of its keeper (see note **A** opposite)

Name Mr/Mrs/Miss ______

Address ______

Postcode ______

Telephone number ______

DETAILS OF THE RECEIVING VEHICLE

(the vehicle which will receive the mark)

Vehicle Registration Mark ______ 9

Make ______ Model ______

VIN/Chassis number ______

If the vehicle is not yet registered, please tick the box ☐

If you are going to post this form, give the expiry date and period of the licence disc if any ______ 19 ___ ; ______ months

Name and address of its keeper (see note **B** opposite)

Name Mr/Mrs/Miss ______

Address ______

Postcode ______

Telephone number ______

Please turn over

DVLC (CV) use only

Donor V5 Doc Ref No. ______

Receiving V5 Doc Ref No. ______

Official Use Only

CD ☐ 3

Replacement Mark ☐ 4 CD ☐ 5

Test Station No. ______

Serial number of Test Certificate (Donor vehicle) ______

RV/GEN II No. ______

V53 if appropriate

Section code ☐ 6 Output marker ☐ 7

Action type ☐ 8 Same keeper

CD ☐ 10 ☐ 11

Licence Serial Numbers

Replacement licence donor ______

Replacement/first licence recipient ______

VRO stamps

Date received	Date processed

PREVIOUS KEEPERS

The Log Book may contain details of the previous Keeper. If it does not, or if additional information is wanted, the new Keeper may obtain details of previous Keepers from Enquiries of the Record Section, DVLA, Swansea, SA99 1AN.

DISPOSING OF A VEHICLE

When selling a vehicle the Registered Keeper must complete the lower part of the Log Book headed 'Notification of Sale or Transfer', detach it and send it at once to the DVLA. Keep a separate note of the buyer's name, address and note the date. Give him the top part of the Log Book and do remind him to inform the DVLA that he is the new Vehicle Keeper; the document states how to do this.

If you are involved in an accident in which your vehicle becomes a total loss and you transfer the vehicle to an insurance company for a total loss payment, you should inform the DVLA that the vehicle has been transferred to the insurance company. Do this by using the 'Notification of Sale or Transfer' part of the Log Book and give the top part of the document to the insurance company.

Above:

If a 'total loss' payment is received from the Insurers for a badly damaged vehicle such as this; the DVLA must be informed that the vehicle has been transferred to the Insurance Company. *Courtesy Kentish Times Newspaper*

Left:

Form V317 explains how to transfer your existing registration mark to another vehicle owned by you.

Right:
If your vehicle ends up like these, you must give the lower part of the Log Book to the scrap dealer, who should inform the DVLA of its demise.

Scrapping a Vehicle

If you have actually broken up or destroyed your vehicle yourself then you must inform the DVLA that the vehicle has been scrapped. You can do this by using the back of the form (section 2) to notify the DVLA.

If you sell or pass the vehicle on to someone else for scrap, even if it is to a scrap-dealer, tell the DVLA of the change of ownership using the tear-off part of the Log Book. Do this as soon as the vehicle changes hands. Hand the top part of the document to the new Keeper so that he can inform the DVLA when he has actually scrapped it.

3 Insuring Your Car

Below:
This new driver is taking the first steps towards finding out about insuring her first car. Always discuss types of insurance and likely costs. Initially premiums are high for novice drivers. Advice can be sought from an Insurance Broker as to what type of insurance suits individual needs.

INTRODUCTION

All owners of motor vehicles used on the roads of the United Kingdom must be insured against the possibility of accidents or personal injuries which may occur to other people (ie. the Third Party) or their property. The law requires that the driver must be insured to drive, either under his or her own policy, or under the policy of the owner of the vehicle.

'Used on the road' not only includes driving, but also leaving your vehicle on the road, even when it is incapable of being driven because it has broken down or is awaiting repair.

Driving without insurance is a very serious offence and will be dealt with quite severely.

You may of course already have an Insurance Agent or Company with which you have been dealing for years on other matters who may be able to offer you a most competitive Motor Policy suitable to your needs. But motor insurance is such a complex business and the terms of policies differ to such an extent that it may pay you to shop around and compare the premiums, conditions and benefits offered. For example, some insurers, when relating to Personal Accidents, make benefits payable from £1,000 up to £25,000 in the event of the death of the policy holder or his spouse (and some policies even include other members of the family) so such benefits should be sought after and carefully considered.

However, if you don't want to shop around and make your own arrangements and you feel you need guidance, then discuss your needs with an accredited Insurance Broker.

ARRIVING AT A PREMIUM

Insurance is a question of utmost good faith on both sides and it is the responsibility of the individual seeking insurance cover to divulge to the insurer at the outset all relevant facts which may affect the insurer's judgement in assessing what premium should be paid, and to advise him of any change in conditions such as ill health, convictions, accidents etc. at subsequent renewals.

SPHERE DRAKE

Proposal Form *Private Car Insurance*

Tick box as appropriate SPHERE DRAKE Policy No.

1. Proposer BLOCK CAPITALS PLEASE

Surname Mr/Mrs/Miss/Ms | First Names

Address | County | Postcode

Town

Are you expecting anyone under the age of 25 years to drive? Tick box Yes No

2. Drivers

Please provide details of YOURSELF AND ALL OTHER DRIVERS who will or may drive the car. DRIVERS UNDER 21 YEARS OF AGE WILL BE EXCLUDED UNLESS NAMED HERE. You may be asked to provide a copy of your driving licence and that of any other driver in the event of a claim.

Name	Date of birth	Married or single	Business or Occupation (including part time)	Nature of your employer's business	Licence: Full, Provisional, E.E.C., International, Foreign. Date driving test passed — Type	Date: Month	Year	No. of years continuously resident in United Kingdom	Main User of the car TICK BOX
The Proposer									

Have you or has any person who may drive — Tick box — Give details if 'Yes' to any part of any question.

(a) had a proposal declined, a policy cancelled, renewal refused or any special terms imposed? Yes No

(b) defective vision or hearing, any mental or physical infirmity, a history of diabetes, epilepsy, fits, any heart complaint or any medical condition? Is any medication being taken in respect of these or any other medical condition? Yes No

(c) been convicted of a motoring offence or any prosecution under consideration? State driver, offence, date, code, and fine. Yes No

(d) been involved in a motor accident, claim, or loss within the last 5 years? State driver, date, cost, details and whether loss was recovered. Yes No

(e) been disqualified from driving? Yes No

If necessary please continue on a separate sheet

3. Car Details

Make & Model	Engine c.c.	Year first registered	Type of body	No. of seats	Estimate of value	Date of purchase	Price paid	Registration Number
					£		£	

(a) Is the car "left hand" drive? Tick box Yes No

(b) Is the car garaged at the above address? Yes No If "No" please give details below, including postcode.

(c) Has the car been altered in any way from the maker's standard specification? Yes No If "Yes" please give details below

(d) Do you own the car? Tick box Yes No If "No", who does?

(e) Is the car registered in your name? Yes No If "No", whose name?

If "No" to (d) or (e) please explain why insurance is to be in your name.

4. Other Cars

If there are any other cars in the household (including company cars) please give details — Make & Model — Owner — Main User — Registration Number

5. Use

Indicate the Class of Use required – see overleaf for definitions. Tick one box only Class 1 Class 2 Class 3

N.B. Make sure you have chosen the correct business use.

6. Cover

Comprehensive / Third Party Fire & Theft / Third Party Only / Third Party Fire & Theft Windscreen

IS DRIVING TO BE RESTRICTED — To you only? / To you and your spouse only? / To persons named in section 2?

Note: Persons under 21 will be excluded unless you name them in section 2 above.

Tick box if "ANY DRIVER" cover required (SPHERE PROPOSERS ONLY AGED 25 OR OVER)

SPHERE PROPOSERS ONLY

Do you wish to bear a Voluntary Accidental Damage Excess? If so tick one box only. £50 £75 £100

N.B. Only available to Proposers 25 or over with Comprehensive cover, and is in addition to any compulsory excess that may apply to your policy.

7. No Claim Discount

Are you claiming any No Claim Discount? Yes No

If "Yes" state number of years entitlement — Years — and ATTACH LAST INSURER'S RENEWAL NOTICE.

If no previous insurance state how driving experience was gained.

Do you require No Claim Discount Protection Benefit? Yes No SPHERE PROPOSERS ONLY - For details see overleaf.

8. Period of Insurance

Insurance to commence on — Tick one box only — For 12 Months / For 6 Months* / For 3 Months*

* DRAKE PROPOSERS ONLY

Important:

Before signing this form, please read again the questions and answers especially if not completed in your own hand. It is an offence under the Road Traffic Act to make any false statement or withhold any material information [illegible] regards as likely to influence the acceptance and assessment of the proposal [illegible] to do so may invalidate your policy. If you are in any doubt about facts which may be considered to be material these should be disclosed for your own protection.

- The Company reserves the right to decline any proposal or to impose special terms.
- The liability of the Company does not commence until a Cover Note or Certificate of Motor Insurance has been delivered to the Proposer.
- A specimen of the policy, giving full details of the standard cover, can be made available for inspection.
- You should keep a record (including copies of letters) of all information supplied to us for the purpose of entering into this contract.
- A copy of this proposal form will be supplied to the Policyholder on request within three months of its completion.

Declaration:

I declare that to the best of my knowledge and belief the above statements completed by me or on my behalf are true and that I have not withheld or concealed anything that might influence acceptance of the insurance. I further warrant that this proposal and declaration shall be the basis of the contract between me and Sphere Drake Insurance p.l.c.

I agree to accept the Company's policy applicable to this insurance.

Signature of Proposer — Date

SD 33/90

Right:

A typical car insurance proposal form on which an applicant is required to give personal and vehicle details so the Insurer can arrive at a premium.

The premium is usually arrived at by taking into consideration such particulars as already disclosed, the driver's age and driving experience, the cubic capacity of the vehicle's engine and the age of the vehicle. The value of the car, particularly if it should be a veteran or a vintage car of great value, will increase the premium.

Although insurance companies differ with regard to the conditions written into their policies the following information generally applies.

The Inexperienced Driver

For a start, a person who has not driven a vehicle for 12 consecutive months is regarded as an inexperienced motorist and is consequently a much higher risk for insurers. This alone means the premium is high for someone just starting their motoring life. Age also comes into the reckoning; anyone under the age of 25 is not likely to obtain a discount, whereas some companies allow a small discount for drivers over 25 years of age, and even more for those over 35 and yet more for mature drivers over 50.

Value And Use

An engine with large cubic capacity or a sport's model with a high top speed will increase the premium. The area in which the vehicle is to be used, such as a city environment or a rural setting, and whether it is to be kept in a garage or not, is also taken into account. The actual value of the car is also significant as repairs or replacements are correspondingly more expensive if any damage is done.

Specialised policies for Veteran or Vintage cars are available which are relatively cheap in comparison to normal cover, but they usually have annual mileage limits and other restrictions.

Rare Cars

Rare cars are independently valued by an expert and an agreed price is arrived at between the owner and the Insurer. The car is usually insured under a Classic Car Policy and the premium loaded as to its use and its mileage. These cars generally appear at Car Rallys and perhaps enjoy an occasional run at week-ends.

Premium Discounts Available

As has already been mentioned, all insurance companies give discounts as monetary incentives to the policy holder to drive carefully and thus not make accident or other claims on the policy.

Left:

Veteran cars such as this Ford Tourer are transported to fetes or rallies to avoid the possibility of an accident and keep their mileage within the policy limits.

Other discounts are given regarding age — already previously mentioned; whether a motorist belongs to a professional body such as doctors, teachers, civil servants etc; or whether the vehicle is to be used solely for social and domestic journeys, and whether you are to be the only driver of your vehicle.

Further Discounts — Voluntary Excess Payments

Most Insurance Companies have a 'Voluntary Excess' scheme whereby if the policy holder is comprehensively covered and is not under 21 years of age, he can further reduce his premium by voluntarily agreeing to pay the first part of the cost of each accident claim to his car; the more he is willing to pay, the bigger the discount. When you have reached your maximum No-Claim Discount (NCD) many companies allow you to protect your NCD by paying an excess payment on your premium whereby, if you are unfortunate enough to have to lodge an accident claim, whether it be deemed your responsiblity for the accident or not, you will not lose your NCD. However, there is a limit on the number of claims which can be made before the NCD is affected.

There may be other discounts available for further voluntary excess payments and it is in your interest to ask about them when seeking insurance cover.

Right:
The ubiquitous Mini. Although it falls in the lowest insurance category, group 1, it will still cost a 17-year-old well over £1,200 to insure comprehensively.

No-Claim Discounts (NCD)

All insurance companies give large discounts — provided no claims are made on the policy — as the driver's experience extends into succeeding and consecutive years.

It is usual for insurance companies to quote the maximum premium and then proceed to subtract from this all the allowable discounts.

No No-Claim Discount, of course, is given for the first year of the policy, but providing a claim has not been made or is pending, the annual renewal premium is generally reduced as follows:

At the first renewal by 30%
At the second consecutive renewal by 40%
At the third consecutive renewal by 50%
At the fourth consecutive renewal by 60% (65% some insurers)

This means that you can enjoy the full 60%-65% NCD after four years of claim-free motoring. Conversely, however, if a claim is made at the time you have earned one of the above stated discounts then your discount at the next renewal may be 20% less or pro rata (this may differ with different insurers) for that renewal.

New Driver Premiums

The following examples are included as a general guide to give the new driver some idea of what insuring his car will cost. The undermentioned premiums are quite basic and are quoted for a small car driven by the policyholder only and is subject to the vehicle being garaged, with no adverse features in respect of either the driver or the vehicle. No other extra features are included.

Premiums quoted for a Mini (Insurance group 1)

For drivers aged	17 years	25 years of age
For Comprehensive Cover	£1,714.90	£902.67
For Third Party Fire & Theft	£901.22	£476.28

You can see what importance Insurance Companies place on a few more years of maturity.

VEHICLE GROUPS

There were seven vehicle groups of insurance (nine with some Insurers) and these had great bearing on the cost of the premium. Group 1 was the lowest and least costly whereas Group 7 was the most expensive.

With effect from 1992 the number of groups was increased to 20, to reflect more accurately the value/risk factor. Previously a BMW 520i and Rolls-Royce were in the same group, the BMW is now Group 14, the Rolls Group 17 or above. Generally speaking the new group is about double the original, old Group 4 is now 7, 8 or 9.

Some cars fare much better, others worse, although many facets will determine which group the insurance premium will fall into, a prime consideration is the size and power of the car.

Left:

This elegant Rolls Royce Silver Spirit II — because of its powerful 6.75-litre engine, its beautifully styled body and its overall cost, inevitably puts it at the top of the insurance range — group 17. But if you can afford £145,000.00 presumably you can afford the insurance.
Courtesy of Rolls-Royce

Below:

Comprehensive insurance covers personal items within the limits quoted in the policy. However, it is sensible to lock such items out of sight.

LEVELS OF INSURANCE COVER

There are three main levels of Insurance Cover:

1. COMPREHENSIVE
2. THIRD PARTY FIRE AND THEFT
3. THIRD PARTY ONLY

Obviously the 'Comprehensive' is the most expensive and 'Third Party Only', offering less cover, is the cheapest.

Of course conditions vary from Insurer to Insurer, but the following shows what conditions are generally covered by the three levels of insurance:

Comprehensive Cover

1. THIRD PARTY LIABILITY (including passengers) is fully covered.
2. LEGAL COSTS incurred with the Underwriters' consent are fully covered. Solicitor's costs defending a traffic offence, arising from an accident if a Third Party claim is possible, are covered. This includes defence against a charge of manslaughter.
3. FIRE OR THEFT giving rise to loss or damage to the insured vehicle.
4. DAMAGE to the insured vehicle.

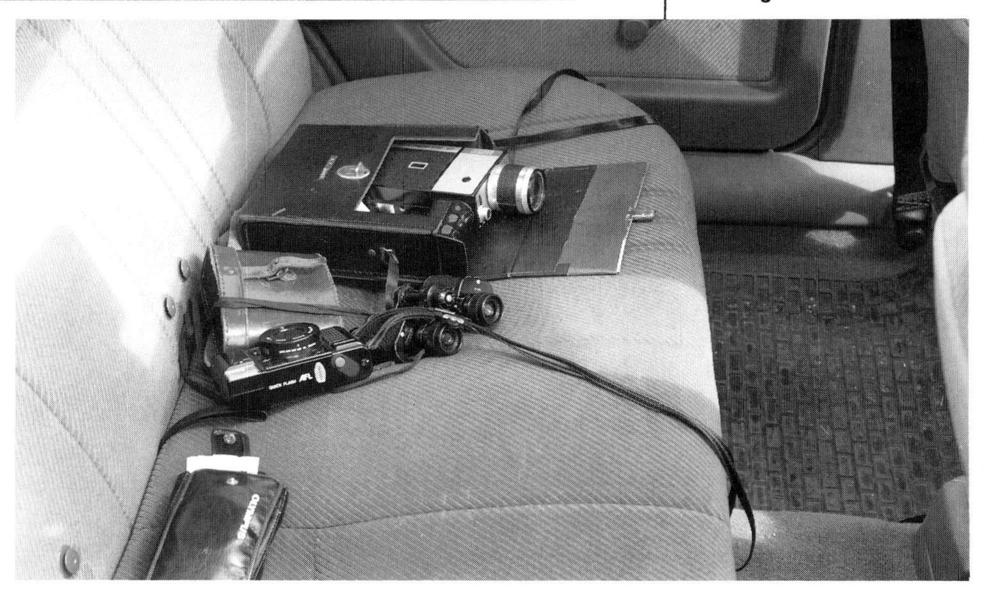

5. REPAIRS may be commenced as soon as a detailed estimate of accident repairs is 'received and agreed' by the Insurers, unless of course the vehicle is beyond economical repair.
6. PERSONAL ACCIDENT COVER to the Insured and/or Spouse, whilst entering, riding in or alighting from the insured car or vehicle not owned by them. Maximum benefit as quoted in the policy.
7. MEDICAL EXPENSES to passengers in the insured car up to the limit quoted in the policy.
8. PERSONAL EFFECTS carried in the car are covered up to the limit quoted in the policy.

Third Party Fire and Theft Cover

1. THIRD PARTY LIABILITY (including passengers) is fully covered.
2. LEGAL COSTS incurred with the Underwriters' consent are fully covered. Solicitor's costs defending a traffic offence, arising from an accident if a Third Party claim is possible, are covered. This includes defence against a charge of manslaughter.
3. FIRE OR THEFT giving rise to loss or damage to the insured vehicle.

Right:
Even in an incident such as this where no one else is involved and it is clearly your fault, your costs should be met if you are comprehensively insured.

Third Party Only Cover

1. THIRD PARTY LIABILITY (including passengers) is fully covered.
2. LEGAL COSTS incurred with the Underwriters' consent are fully covered. Solicitor's costs defending a traffic offence, arising from an accident if a Third Party claim is possible, are covered. This includes defence against a charge of manslaughter.

AFTER THE ACCIDENT POLICY COVER

Even if a motorist has had a successful accident claim met he can find himself considerably out of pocket. The insurance companies have come to recognise this and now offer an 'after the accident' policy which can be taken out even after a considerable time has elapsed since the original claim.

The company will send you a proposal form requesting relevant details of the accident, your motor insurer's accident report, your own losses or damage, whether you sustained injuries and details of any other person you wish to benefit under the policy.

From the information you have disclosed on the proposal form the insurer will indicate whether they think you have a sound claim and will say whether they are willing to act for you.

If your proposal is accepted the initial charge is £25 for yourself and £25 for any other person named on your proposal form. This will be returned to you if the company is not able to quote terms, or if you choose not to accept the terms offered, within the period stated in the quotation.

When a recovery is obtained, a further element of premium, the adjustment premium, will be charged. This is calculated as a percentage, normally between 5% and 20% of the amount recovered. This premium will reflect the size and potential risk and cost of handling the claim, and their quotation will clearly show the percentage to be applied.

If no recovery is obtained, there will be no adjustment premium for you to pay.

This type of policy should particularly appeal to those who have Third Party Fire and Theft cover and find that they are responsible for arranging their own repairs and claiming back the costs or compensation for injury, from the other party.

INSURANCE BROKER

To obtain guidance as to what level of insurance is best for you, consult an Insurance Broker. Brokers do not usually charge for their

AFTER THE
ACCIDENT
POLICY

(THIS POLICY IS DESIGNED TO
BE TAKEN OUT **FOLLOWING**
AN ACCIDENT)

Issued by:

MLP MOTORISTS LEGAL PROTECTION LTD
62-72 VICTORIA STREET
ST. ALBANS
HERTFORDSHIRE AL1 3XH

TELEPHONE 0727 69152
FACSIMILE 0727 61206

Left:
This type of policy can be taken out if, after having been involved in an accident and your normal insurance claim has been settled, you find that you are still out of pocket.

Right:
A car which has been specially adapted to suit a driver's disability should not attract additional loading on an insurance premium.

Below:
This MAVIS pamphlet available from Department of Transport gives invaluable information for the disabled driver.

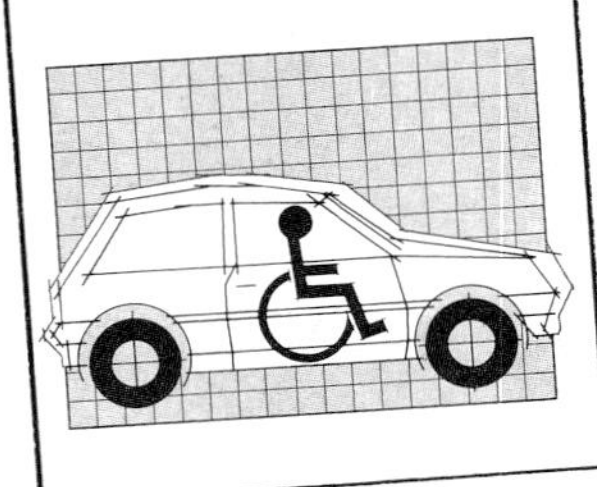

services, but should they wish to do so perhaps for additional work not directly concerned with your insurance, then they will agree a fee with you in advance of the work being done.

Impartial Advice

Brokers are experts in all types of insurance and will be able to tell you what is available and at what cost. Because Brokers do not represent any particular insurance company they should be of independent mind and can suggest one of a number of insurance companies with which they deal. They will be able to suggest what type of insurance to have and how much cover will be needed. They will also be able to help with any claims and will inform you when renewals are due.

If you select an Accredited Broker, that is one who is a member of the British Insurance and Investment Brokers Association (BIIBA) then any dealings you have with them will have the protection of the Association behind it.

INSURANCE FOR DISABLED DRIVERS

The disabled driver must declare the full facts about their disability and describe fully the adaptations made to their car. Failure to do so may invalidate any subsequent insurance claim.

The insurance premium for a disabled person is arrived at by considering all the usual variables: the type of car — whether high-powered or sporty, whether there are any earned No-Claim Discounts to be considered, the area in which the policy holder lives and the age of the driver. New young drivers will find the premiums high, as is the case with any new young driver.

Obtaining Cover

Generally, depending on the scope of the policy and whether Third Party or Comprehensive insurance is required, a disabled person should encounter no problems in getting cover for a properly modified car and there should be no additional loading imposed because of a disability. The cover should include the usual allowance of up to a 60% No-Claim Discount, to be earned for safe driving, and all the other inducements that are generally offered.

Other Variables

Some companies also offer personal accident policies for disabled drivers as a supplementary cover to their motor insurance. Others include schemes to get you home if you should suffer a mechanical breakdown.

It is not uncommon in quotations from some companies to include a premium-loading on a driver for even a mild disability which is easily overcome by adaptation of their vehicle. If you should encounter this type of company, have nothing to do with them.

Shop around and get a number of quotations before making a final decision and do consider carefully that the cover offered is adequate. Check the policy thoroughly and read the small print.

A wealth of information can be obtained from the Department of Transport's Mobility Advice and Vehicle Information Service, referred to in Chapter 1.

SUMMARY

To sum up, a newcomer to motoring can expect to pay a high premium initially, with the premium coming down with consecutive claim-free years of driving.

You will need to decide what level of insurance you require, what discounts are available and whether you want to do the information-seeking yourself or whether to use the services of an Insurance Broker.

It is incumbent on you to make sure that your insurance does not lapse, although most insurance companies send out reminders.

4 Licensing Your Car

INTRODUCTION

A vehicle that is driven, parked or left unused on a public road must be licensed. It must also display its licence (tax disc) on the left-hand side of the windscreen (viewed from inside the car). This shows that the necessary vehicle excise duty has been paid: the tax disc also shows the vehicle's registration mark and the date on which the licence expires. You cannot transfer a licence from one vehicle to another.

Licences may be bought for six or twelve calendar months, although two six-month licences will cost about 10% more than a twelve month licence, due to administrative costs.

To help those motorists who do not wish to or cannot pay for a licence all at once, a Vehicle Licence Stamp Card (form V218) can be obtained from any post office. This takes stamps in £5 units from any post office which are used as part payment for a vehicle licence.

Remember, the only time that an unlicensed vehicle may be driven on the road, providing that it is insured, is to or from a Department of Transport Testing Station for a pre-arranged compulsory Vehicle Test (MOT) see Chapter 10.

OBTAINING A VEHICLE LICENCE

If you have bought a new car from a dealer he will usually arrange for the vehicle to be licensed at the same time as he is applying for the Vehicle Registration Document.

However, if you have bought privately and the car is due to be licensed but sufficient time has not elapsed for the DVLA to have registered you as the new owner you will need to obtain a Department of Transport Form V10 from any Vehicle Licensing Post Office (main post office) or from a Vehicle Registration Office (VRO).

Normally the DVLA will remind you on form V11 two to three weeks beforehand that the licence for your vehicle is due for renewal.

Below:
A tax disc must be permanently displayed on the passenger side of the windscreen. It is illegal to drive a car on the road without displaying one.

An application for a vehicle licence, obtainable for a period of six or 12 months can be made at a Vehicle Licensing Post Office. You will need the following:

1. A correctly completed application Form V10
2. The Vehicle's Registration Document (Log Book). If this is not available complete a Form V62 from a Vehicle Licensing Post Office.
3. A valid certificate of Insurance.
4. A valid Vehicle Test Certificate MOT if one is needed (see chapter 10).
5. The appropriate Duty payable.

If this is all in order you will be issued a licence (tax disc) over the counter.

Whenever the DVLA reminds a Keeper that his car licence is due for renewal the above procedure is followed, except that the Vehicle Registration Document need not be shown at the post office, providing the renewal form V11 sent by the DVLA is used in place of form V10.

A Concession

If for some reason you are unable to renew your current licence at the date due and it has just expired, there is a concession allowing 14 days grace in which you can still use your car providing you have already applied for a new licence to run from the day after the date of expiry of the old licence. This is only a concession and 'in law' a current licence must be displayed at all times when the vehicle is on the road.

If you do not want to apply at a Vehicle Licensing Post Office you can post your application form to the head postmaster (MVL Duty) at a head post office. A list of these throughout the country is given in leaflet V100, obtainable from any Vehicle Licensing Post Office.

TO RENEW A LICENCE

You can renew a licence within the 14 days prior to the expiry date, but not before. This can be done at a Vehicle Licensing Post Office or by post to the head postmaster (MVL Duty) at a head post office providing all the conditions aforementioned are complied with. These are listed in leaflet V100.

DVLA Swansea usually sends the registered owner a reminder (form V11) for Vehicle Licence Renewal about 14 days before the current one is due to expire.

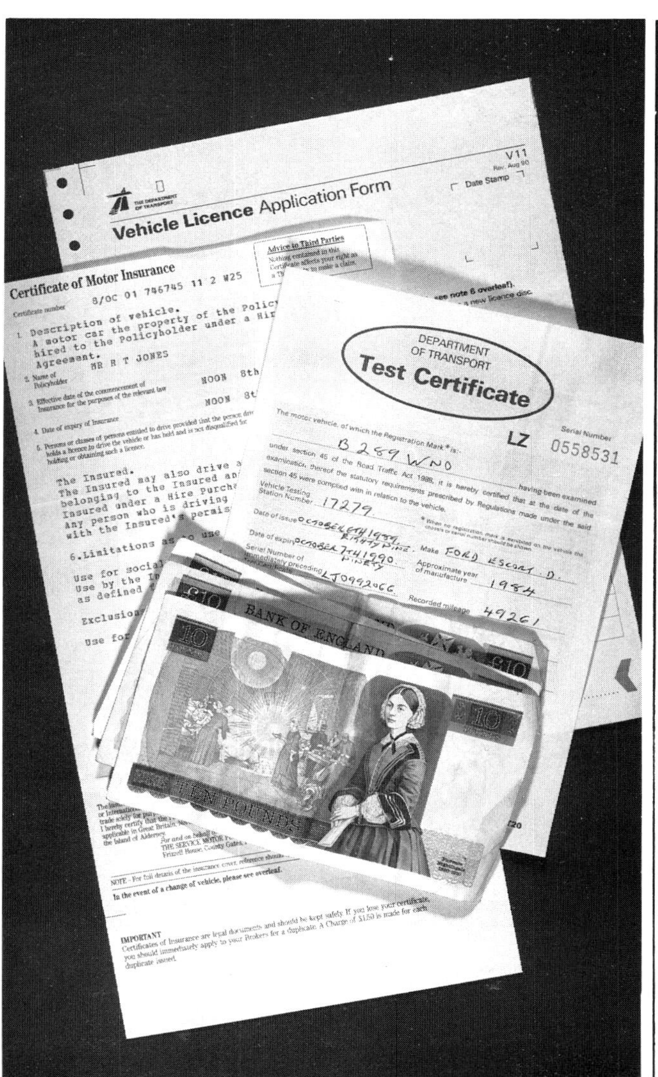

Far left:

The stamp card (form V218) enables the motorist to save towards the cost of his tax by buying stamps in £5.00 units.

Centre left:

If your car has been recently purchased and the DVLA has not yet had time to register you as the new owner, these forms should be presented to any main post office in order to obtain a licence.

Left:

To obtain a concurrent licence at a main post office, these items are normally required with the correct fee.

Duplicate Vehicle Licence Disc . . . application form (V20)

Please read these notes before you fill in this form

When to apply

You will need to apply for a duplicate vehicle licence (tax disc) if:

- your licence has been lost, stolen, destroyed or spoilt in any way
- the colour has faded or the figures on it cannot be read

If you have still got the licence, you must enclose it with this form

If you apply for a duplicate licence because you no longer have the origin later get it back, you must send the original licence to the Driver and Vehicl Centre (DVLC), Swansea SA99 1AG. You should keep the **duplicate** lic vehicle. Duplicate licences are clearly marked "Duplicate". Rememb duplicate licence is issued, the original licence is of no effect and it **must in any way.**

How to apply

If you have a Registration Document (V5) for the vehicle, p application form to a Vehicle Registration Office (their addresses a book under "Transport, Department of"). The duplicate licence will Driver and Vehicle Licensing Centre, Swansea. Provided the reco vehicle is properly licensed. The Vehicle Registration Office will licence for you to use on the vehicle until you get the duplicate lic

If you do not have a Registration Document (V5) for the complete form V62 and post it together with this form to DVLC, Please note: if you apply to DVLC, you will **not** be sent a tempo will only get the duplicate.

What to pay

The fee for the duplicate licence is £3.50. This is the charge cannot be refunded even if you later get the original licenc

How to pay

You can pay the fee by

- Cheque or postal order — made payable to Depart
- National Girobank Transfer
- Cash — if you must send cash, please use regis

V14

Refund application

. . . on return of a vehicle licence disc

REFUND

To help you

Only use this form if you are returning a licence disc for a refund. If you don't have the disc, you must use form V33 instead. You can get a V33 from a Vehicle Registration Office (VRO). The address is in the telephone directory under "Transport, Department of".

You can apply for a refund if you bought the licence disc or if it was passed to you with the vehicle.

When you have completed and signed this form, send it with the licence disc to Refund Section, DVLC, Swansea SA99 1AL or hand it in at any VRO.

When to apply

You get a refund for each **full** month still to run on the licence disc at the time you apply for the refund. To get a refund for any one month, the licence must be —

- posted to DVLC **before** the first day of the month, or
- handed in to a VRO **before** the first day of the month

Remember, if you bought a 6 monthly licence disc, you paid an extra handling charge of 10%. This isn't included in the refund.

Motor dealers, fleet owners, etc

Do not use photocopies of application forms. Payments may be delayed if you do.

If you use rubber office stamps for the name and address section, please make sure that the names and addresses are shown clearly.

Disabled people

If you have been granted exemption from vehicle excise duty, and are claiming a backdated refund, please also complete form V285, which you can get from any VRO. You will need to take or send the completed forms V14 and V285, together with your exemption certificate, to your nearest VRO.

If you have sold or scrapped the vehicle, you must tell DVLC. Return your Registration Document or write, giving the registration mark, the date of sale or scrapping and the name and address of the person who now has the vehicle. This will make sure you are no longer shown as the keeper of the vehicle.

TO REPLACE A MISSING LICENCE

If your vehicle licence is stolen, destroyed, lost, mutilated or defaced, or has become illegible, you can apply for a duplicate, using form V20 which can be obtained from any Vehicle Licensing Post Office or any VRO.

Your application form together with the fee (form V20 states the necessary amount) and the Vehicle Registration Document, should either be taken to or be sent to a VRO. If the vehicle is registered in your name and a licence is still in force you will be issued with a duplicate covering the full licence period.

It may be possible to have issued a duplicate licence free-of-charge where the Secretary of State for Transport is satisfied that the figures or particulars have become illegible through no fault of the applicant.

OBTAINING A REFUND ON AN UNEXPIRED LICENCE

It is possible to claim for a refund on an unexpired licence for a period of one month or more. Refunds can only be made for complete calendar months and the application must be received before the first day of that month.

An application for a refund should be made on form V14, obtainable from Vehicle Registration Offices (VRO's) or from Vehicle Licensing Post Offices. These, when correctly filled in, should be posted to Refund Section, DVLA, Swansea SA99 1AL. Allow up to six weeks before making enquiries.

You may apply for a refund if your licence disc has been lost or destroyed, or if your vehicle has been stolen. A special application form (V33) is available for this purpose from any VRO.

Far left:

If your original licence goes missing for any reason this form (V20) enables you to obtain a duplicate.

Left:

To claim a refund for any unexpired portion of your licence use this form (V14).

5 Special Driving Conditions and Techniques

INTRODUCTION

When they were first founded both the RAC and AA motorcycle patrolmen saluted every member of their respective organisations whenever they were encountered on the road. Of course we cannot expect this nowadays as there are something like 20 million motorists on the road, but courtesy between motorists should be maintained.

Some of the main causes of annoyance on the road are deliberately blocking another lane of traffic if you yourself cannot move; not letting traffic from a side road filter in; lane hopping in an attempt to get ahead in a jam and a general lack of consideration and manners towards other road users.

Fitness to Drive

If you are about to tackle a long journey make sure you have rested adequately beforehand. It is irresponsible to undertake a long journey, particularly on a motorway, if you are already feeling jaded. Driving on a motorway where there are no road junctions, no traffic lights and no gear changes to relieve the monotony can quite easily cause drowsiness. Falling asleep at the wheel is not uncommon and extremely dangerous. The police say that 25% of all accidents on the M1 are due to this.

If you feel the slightest bit drowsy, stop the car (not on the hard shoulder of course), stretch your legs and do deep breathing exercises. This should be repeated every one and a quarter hours or

Left:

The traffic proceeding to the left along the main road was halted at traffic lights. The Royal Mail van should have left a gap to assist the cars waiting to turn right.

Right:
An accident such as this in which no one else was involved, occurring late at night, suggests the driver had fallen asleep – with terrible consequences.

Below:
This motorist, quite rightly, is taking time to thoroughly clear the early morning frost from all windows before commencing her journey. Driving with frost on the side windows is inconsiderate as other drivers need to see through the car.

so. Stop for refreshments every 2½ hours, but beware if you have a heavy meal, because the incidence of tiredness will increase and always refrain from drinking alcohol whilst driving.

Clear vision

If a car has been parked out in the open throughout a night of frost do allow time to clear all windows before setting off. One often sees a driver peering through a small oval space that he has cleared from his windscreen. Apart from being dangerous, this is against the law.

Apparently the British are notoriously neglectful about keeping windscreens, windows, lamps, mirrors, etc clean. Although we drive some of the best equipped cars in Europe, we often drive with the worst windscreen vision. It is absolutely essential that your all-round vision is clear; not only for you but for other motorists who rely on being able to see through your car so that they can react promptly in time to any situation.

Road Conditions

Unfortunately, when taking driving lessons the learner does not always have the opportunity to drive in all types of climatic conditions and it might be prudent to cover some of these here. Although some of the following points are mentioned in *The Highway Code*, it is worth repeating them.

Speed limits are set for the average motorist in the average car in good weather and road conditions. When any of these factors vary the speed limits may be too high.

Good driving, ensuring absolute safety, requires 100% concentration, judgement and experience of what you, your car and the road conditions will allow at any given time. Mirrors should be used every few moments as the road situation changes constantly.

Fundamentally, all braking, gear changing and accelerating should be done on the straight before one begins to turn into a bend or when cornering. Skids are usually caused by excessive speed, acceleration, or braking on a bend or corner.

It is not always fully appreciated by the motorist that the area which holds the car on the road, or more correctly 'to the road', is little more than the size of the sole of a man's shoe — for each tyre. While this is adequate most of the time, the adhesion of the car to the road becomes significantly less in wet, muddy, icy or snowy conditions and particularly when cornering. It is as well to remember this fact as it should ensure that you always drive carefully at an appropriate speed for the road conditions.

Centre left:

This driver does not have adequate vision and is therefore behaving irresponsibly.

Left:

It is essential to have clear vision and these wiper blades quite obviously need replacing. This vehicle would fail its MOT test on this point.

MOTORWAY DRIVING

If you are new to motoring you will have little or no experience of motorway driving. Motorways are the safest type of road as they have no road junctions, sharp bends, roundabouts, steep hills or traffic lights to negotiate and all the traffic travels in the same direction.

Before venturing on a motorway particularly if you are considering a long journey, you should feel fresh and alert. The vehicle should be checked, the tyres inflated to their correct pressures — or even raised a couple of pounds (lb/in^2 – or an extra 0.1 Bar) if so recommended by the manufacturer — and see that there is adequate fuel, oil and water being carried. Running out of any of these could be dangerous, costly and inconvenient.

Be Aware And Alert

In good, dry, clear visibility you should allow a minimum distance from another vehicle in the same lane of at least 1 metre per mile of speed of which you are travelling ie: 60 metres for 60mph. For wet or icy conditions you could need 10 times more than this.

You must be aware that, because of the greater speed at which vehicles are travelling, traffic patterns change much more quickly than when motoring on other roads.

Above:
Clearing the windscreen with a plastic scraper after a heavy frost, will not damage the windscreen.

Right:
This acceleration lane is of ample length and width to enable the motorist to attain the speed of the motorway traffic before joining it.

Left:

The two cars on the extreme left have quite correctly used most of the acceleration lane to attain the speed of the motorway traffic.

Right:
When overtaking, having established it is safe to do so, signal early, build up your speed, move out well back from the vehicle you are overtaking and complete your manoeuvre speedily before returning to your correct lane.

Right:
Early warning that the motorway divides into two. Pay attention to the overhead direction signs as you may need to change lane more than once.

Far right:
The large signs giving warning of a road junction one mile ahead show only road numbers.

Use your mirrors more frequently and always be on the look-out for cars approaching at speed to overtake.
Any manoeuvres you wish to make should be done in good time with adequate signalling.

Entering A Motorway

Access to most motorways is via a roundabout that leads into a slip road, the acceleration lane.
Use this to accelerate and adjust your speed so that you can filter in when it is safe to do so.
Stay in this lane until you have acclimatised yourself to the speed and the flow of traffic, and always give way to traffic already on the motorway.

Changing Lanes

Change lanes only when there is a need to do so. When overtaking, bear in mind how quickly vehicles are coming up behind you. Use your mirrors, signal your intention early and complete your manoeuvre quickly. Start the manoeuvre early to ensure that you have sufficient acceleration time to overtake quickly. When returning to your lane don't cut in too early or too close to the vehicle you have just overtaken.
See that you cancel your indicators as the comparatively slight movement of the steering wheel may not be sufficient to operate the self-cancelling device.
Remember, the right-hand lane is not the fast lane, it is an overtaking lane, so don't stay in it longer than is necessary. Caravans, trailers of any type or heavy goods lorries are not allowed to use the right-hand lane.
Where motorways join or separate you may need to change lanes and sometimes more than one lane change is necessary. Pay attention to the overhead direction signs and move into the correct lane in good time.

Leaving a Motorway

Motorway exits are generally signposted one mile in advance.
The first sign gives road numbers, then in ½-mile another sign gives place-names as well. As you approach an exit, count-down markers are placed 300 metres, 200 metres and lastly 100 metres before the deceleration lane begins. Use this succession of indicators to change your lane if necessary and to reduce your speed.
Remember, that after travelling some miles at speed it is easy to underestimate how fast you are going — 50 or 45mph may seem more like 20mph, so glance at your speedometer to ascertain your true speed.

Paddock Wood
A 20 (B 2016)
Gravesend
Tonbridge
(A 227)
Wrotham
2
½m
Maidstone
Folkestone
Dover
Paddock Wood
A 20 (B 2016)
Gravesend
Tonbridge
(A 227)
Wrotham

Far left:
The warning signs of a junction half a mile away show place names and road numbers.

Centre left:
The 100m count-down marker is the last reminder that you are nearing the deceleration lane for leaving the motorway.

Left:
Driving in wet conditions is quite hazardous as oncoming lights reflect off the wet road surface and shop windows.

Your Safety

All the aforesaid applies to good climatic conditions. If you encounter rain, ice, high winds or fog reduce your speed and leave greater distances between other vehicles. Fog particularly affects judgement of speed and distance, so drive cautiously.

NIGHT DRIVING

When driving at night make sure that your all-round vision is clear and that your lights are in working order and properly adjusted so as not to dazzle oncoming motorists. Remember, by day or night, regardless of conditions, you must be able to stop within the distance you can see to be clear. Look out carefully for cyclists and motorcyclists who present a small profile. Cycles are frequently poorly lit.

In a built-up area at night you should drive with more diligence than might be called for in daylight and take great care where there are patches of light and shadow as pedestrians are often difficult to see under these conditions.

You must use headlights on all roads where there is no street lighting.

Headlight dazzle

If you are dazzled by approaching headlights, slow down, keep well to the left and even stop if necessary.

Always dip your headlights when other traffic is approaching and when following another vehicle if you are close enough for your lights to dazzle the driver.

If you are troubled by glare reflecting in your driving mirror move your head or body so that you are out of the range of the reflected lights. A dipping mirror, if fitted, can of course be operated to avoid being dazzled from following vehicles. Beware of wet conditions where lights are reflected from all angles, making conditions difficult for the driver.

DRIVING IN FOG

When driving in fog or mists, whether it be in darkness or daylight, all drivers *must* use their headlights (Highway Code), or foglamps if the vehicle is so equipped. This affords one greater protection by ensuring earlier recognition from approaching drivers. Unfortunately it is not uncommon to see cars being driven with sidelamps only, in mist or fog. Check and clean windscreens, lamp reflectors and windows whenever you can. Remember that in fog or mist steamed-up windows can further reduce your visibility, by as much as 30% to 50%.

Keeping Your Distance

When following another vehicle keep your distance comparable to the speed at which you are travelling so that you can pull-up with safety if suddenly required to do so. Watch your speed when travelling on what appears to be a deserted road, there may be unexpected hazards ahead.

Do not speed up to get away from a vehicle whilst it is travelling too close behind you.

Use white-line markings and cats-eyes as a guide only, and rely on your own eyesight.

Making a Turn

Turning at a right-hand junction in fog needs a lot of care. Open your window so that sound can aid you and start indicating early. If you have to wait for other vehicles to pass before you can turn, keep your foot on the brake pedal so that your stop lights are on as an extra warning to following traffic. Use your horn and listen for possible replies. Don't turn until you are as sure as you can be that it is safe to complete your turn.

If visibility is seriously reduced, say to 100 metres or less, the rear high-intensity foglamps should be switched *on* as well. These should only be used when visibility is very poor, however.

CADENCE BRAKING

You may have heard the expression 'Cadence Braking' and wondered what is meant by this. Well, in order to stop a car as quickly as possible, sufficient brake pedal pressure should be applied to *almost* lock the wheels. If the wheels are completely locked then a skid or slide will develop and this will actually result in loss of braking power.

In cadence braking, having applied pressure to the brake pedal, it is then lightened just before locking of the wheels would occur

Far left:

Although it is not actually very far away, the second oncoming car in the near-side lane can hardly be seen as it does not have its lights on. The headlights of the fourth vehicle on the other hand can clearly be seen – thus illustrating how important it is to use headlights in poor visibility.

Left:

A good example of motorists driving at a sensible distance from each other; although the fog is not dense here, remember you may suddenly hit a dense bank of fog at any time.

and then the pressure is re-applied; this is repeated several times extremely rapidly enabling the car to stop quickly but safely.

This method is particularly useful when trying to stop on wet or icy roads. It is important to avoid locking the wheels, yet not let the car 'run loose' when releasing the brake pedal.

Practising The Technique

In fact more and more new cars are being fitted with anti-lock braking systems so this method of braking can only be applied to older vehicles. However, if you do get the opportunity to practise on a deserted piece of land such as an old unused road or an abandoned airfield — then try it out — but do exercise caution. If you can master the technique you will find it will increase your control and confidence when handling a car.

DRIVING IN SNOW

The best advice is 'not to travel' if expecting snow, but if you are already miles from home or, in spite of the weather, you have to make a trip, then being 'prepared for the worst' is the next best policy.

You should make a point of carrying chocolate, sandwiches and a flask of hot drink, wellington boots, heavy coats, gloves, blankets, snow chains and a shovel and possibly some grit or ballast in the boot. Grip mats with long strings are also a good idea.

There are still remote parts of the United Kingdom where these items could mean the difference between life and death.

If you do become stranded in a remote area your best course of action is to stay with your car. Setting off on foot may spell disaster (see paragraph In Deep Snow, page 63).

It is possible to drive in snow to a depth of 6 inches, but anything deeper than this makes progress extremely difficult, if not impossible.

Constant Steady Speed

When driving in snow you should endeavour to maintain a speed which requires very little gear changing and, if at all possible, no stopping. Third gear is a good gear to operate in; it is sufficient to pull the vehicle at low speed without stalling the engine and it is capable of allowing you to accelerate to a reasonable speed if the traffic flow permits. The point of not changing gear too often is to lessen the tendency to lose traction and induce wheelspin, or skidding.

By astute driving you can often avoid stopping and this is a great boon if you are to progress in snow. If you do have to pull away from a stop, be very light on the throttle; you need just enough throttle to turn the driving wheels without creating wheelspin.

Use of gears and throttle

If conditions are really bad it often helps to select a higher gear — say second — for pulling away from a stop as this tends to reduce wheelspin.

If driving a car fitted with an automatic gearbox lock the gear by selecting position 1 or 2, depending on the design of your particular transmission system.

Keep in as high a gear and as small and gentle a throttle opening as is possible. Selecting too low a gear and being too heavy on the throttle will race the engine and inevitably cause wheelspin.

Even some experienced motorists seem to be under the erroneous belief that if you get wheelspin you need more throttle to

Far left:

This driver is trying to improve his braking technique by practising the cadence method, on an abandoned road.

Left:

This driver is removing her snow boots to don sensible footwear, which should be worn at all times.

Left:

Be especially careful of pedestrians walking in the gutter and give them as wide a berth as possible.

compensate for this. This is utterly wrong, you only get yourself deeper into trouble by doing this.

Gentle Braking

Although it is best to avoid stopping it is inevitable that this will be necessary sooner or later.

To do so, apply the brakes very gently to avoid skidding. Changing down through the gears to brake the vehicle may in fact induce skidding.

Stuck On A Hill

If you do get stuck you may be able to get out of trouble by shovelling the snow away from the driving wheels for a distance of a few yards — say nearly the car's length — so that when you do get traction you can keep going.

When you have cleared a path, put the grit down right up to the tyres. Release the handbrake and allow the car to roll back on to the grit.

Above:
Getting going is not easy on mornings like this. Snow deeper than this would pose real difficulties.

Right:
Note how the rear of the leading car has skidded to the left as the driver accelerated away after completing his turn.

Left:

These cars are proceeding with caution as they are negotiating a long descent; the rear driver is braking gently to stop his vehicle gaining on the other. It would be prudent to leave a greater gap between the two cars.

Remember to put the shovel and grit back in the boot before attempting a start because once you are on the move you will not want to stop.

Rocking Out of Trouble.

In really difficult cases it might be necessary to 'rock' the car. This is achieved by selecting a low gear and then with a slight amount of throttle trying to pull away, ie. putting a load on the driving wheels then slipping the clutch and alternatively driving and slipping the clutch. This tends to rock the car, the idea being to take avantage of each little forward movement until a positive forward momentum is achieved. It is then a matter of keeping this going until the car can be driven forward. With an automatic gearbox it is a matter of selecting D or R quickly to achieve the same results. This means

Right:
The driver made the mistake of changing gear on a steep hill and now finds that he cannot progress any further because his driving wheels are spinning. The solution is to clear the snow from the wheel tracks, allow the car to roll back a car's length; select a low gear, throttle gently and pull away steadily without accelerating until the summit is reached.

selecting D when the car is actually moving backwards and R when the car is moving forwards, but it is safe to do so as long as a small throttle opening keeps the engine speed low.

Snow Chains

In extremely heavy snow conditions increased traction may be obtained by fitting snow chains, providing your tyres can take them.

Ideally, all the wheels should be fitted with chains but if only one pair is available they should be fitted to the wheels on the same axle and on the driving axle at that.

It is advisable to use chains with small links to ensure sufficient clearance between tyre and wheel arches.

Remove the chains when not required.

Driving Up A Hill

When approaching a snow-covered hill, particularly a steep one, you will need to quickly assess which gear will take you all the way up. Get into this gear (usually first or second) as you approach and keep to a constant speed until the summit is reached. Even though you know that you could get into a higher gear and go faster up the hill resist the temptation because it is the action of changing gear,

Right:
This motorist, quite sensibly, is fitting snow-chains to his driving wheels — absolutely necessary to prevent wheel spin in this case as he is towing a caravan. Easy-fit snow chains such as these make driving in snow conditions much easier and safer.

Left:

The car is about to climb a hill so it would be prudent to stay in a low gear and proceed steadily without accelerating until safely at the top.

Right:
This competent driver has proceeded down the hill in a very low gear without letting his vehicle's speed increase and now that he needs to stop for the passing lorry, only a gentle application of the brake is needed.

when you lose a little bit of momentum, and then trying to snatch it up with some acceleration which causes skidding and loss of traction. This must be avoided at all cost on hills.

Descending A Hill

Descending a snow covered or icy hill is extremely tricky and must be approached very cautiously. The best action to take is to change to bottom gear early — while still on the summit. Do not allow the car to increase speed — not easy I know — but with the engine ticking over and in bottom gear and with perhaps a very light early application of the brakes — not enough to lock the wheels, but sufficient to keep the vehicle from over-running — you should be able to negotiate a safe path downhill.

The object is to get the engine via the low gear to do most of the vehicle braking so that the wheels merely roll slowly down the hill without gaining momentum and needing to be braked – certainly not harshly. On automatics use the lock-up or hold to maintain a low gear.

Remember when descending a steep hill it is courteous to give way to vehicles coming up the hill.

In Deep Snow

Hopefully you will never be trapped in deep snow, but if you do find you are stuck and experiencing blizzard conditions and drifting, don't leave your vehicle, unless a house is in sight.

Fully extend the aerial and tie a marker to it, or put the spare wheel on the car roof. Settle down with your hot drink (unmelted snow should never be used as a substitute for a drink), blankets and clothes and make yourself comfortable to await rescue by the local authorities who will be out looking for people who are trapped.

Don't run the engine unless you know your exhaust is free from snow. If blocked, the fumes may get into the car with dire results. *Exhaust fumes can kill.* Open a window away from the wind occasionally to get some fresh air.

SKIDDING

Skidding is caused by the way a vehicle is driven. It may be that it was driven at too great a speed, it was put into a too-tight turn or it was braked too harshly. There are also other contributory causes such as poor tyres, wet or icy road conditions or poor suspension. Even so, it is still possible to make a vehicle skid with new tyres, a dry road and excellent suspension.

Above:
These vehicles were checked-out by the authorities but the occupants had vacated them and presumably sought sanctuary earlier in the snow storm.

Left:
In making a turn to the left the rear of the car has skidded to the right. The driver has instantly had to turn his steering wheel to the right to neutralize the skid, and then to straighten up in the next few feet.

Right:

Skidding is all too easy and frightening in arctic conditions, Here the cars are being driven correctly as there is sufficient space between them to avoid braking harshly and thus risking the possibility of skidding.

However, what action do we take having got into a skid? Unfortunately skid control is not taught to learner drivers and for many people their first skid will be a completely unexpected experience. The most important thing is not to panic.

Controlling the Skid

Let us assume you are driving round a left-hand corner when your car goes into a skid. Immediately you are aware that the car is not responding to your steering — you have lost control of it and the rear of the car is sliding in an anti-clockwise direction as if it is pivoting about the front wheels. Take your foot off the accelerator pedal immediately and at the same time apply steering correction for a very short time very quickly to the right, as if to do a right hand turn ie. in a clockwise direction.

This all has to be done in an *instant* for it to have the effect of neutralizing the skid and straightening up the path of the vehicle within a matter of a few feet. This is what is meant by saying, 'driving into the skid'.

Excess or prolonged steering correction should be avoided or another skid may be induced in the opposite direction.

Avoiding Skids

Try and avoid skids by driving in such a way as to prevent them.

Winter of course is the most testing time and stopping distances may need to be up to ten times greater than during a dry period. Be on the look-out for slippery roads. The sight of snow is an obvious danger, but on many occasions, perhaps on a dark wintry night or a frosty morning, you might be taken completely unawares by hitting a spot of black ice on an exposed junction or open part of the road. Be aware and alert at all times in wintry conditions. Having said that, skidding can occur any time — in the wet, on wet leaves, muddy roads or even on loose shale, so the watchword is always be aware of the condition of the road surface and keep to a speed consistent with the prevailing conditions.

DRIVING THROUGH HEAVY RAIN

When driving at speed in heavy rain be extra careful if the water is not draining quickly from the road surface because water-logged tyres may lead to aqua-planing particularly if the front tyre treads are getting worn.

Above:

There is a very real likelihood of black-ice on this exposed roundabout. Extreme caution will be needed to negotiate it.

Left:

This driver should have slowed down because by travelling at this speed he is likely to induce aqua-planing.

Aqua-planing occurs at speed when the front tyres cannot cope with the undispersed volume of surface water. The tyres push superfluous water forward, creating a wedge effect and, given time and distance, the tyres will mount the wedge and lose contact with the road.

The symptoms of aqua-planing are light steering, and, if speed is maintained, complete loss of steering control. If aqua-planing is suspected one should instantly come off the accelerator pedal, do not brake, but allow the vehicle to lose speed. When the speed has reduced considerably the tyres will make contact with the road again and steering control can be regained.

Too many drivers ignore adverse weather conditions and maintain excessive speeds even when it is difficult to assess water depth. Another hazard is that the water could be hiding deep ruts or pot-holes and if you are travelling too fast the steering could be wrenched out of your grasp.

If you have been travelling for some while in heavy rain it is expedient to try your brakes at a convenient time to reassure yourself that they are still working effectively.

Trunk Roads

When travelling on trunk roads in heavy rain be prepared to be overtaken by heavy vehicles which may throw a drenching heavy mist of spray over your car, completely obscuring your vision for a moment or two.

This is extremely hazardous and makes it difficult to detect further surface water, thus creating potential danger for yourself and other road users. The best action to adopt is to be alert to the situation, be constantly looking in your mirrors so that you are not caught unawares and make sure you are looking well ahead all the time to see that your path is clear while being overtaken.

Right:
This flash-flood formed on a motorway after a 30-minute downpour. If you should encounter anything similar, drive carefully and slowly so as not to create a bow-wave as this lorry is doing and dry your brakes when clear of the water.
Courtesy Daily Mail

Facing page, top:
Overtaking this long vehicle from the centre lane should not be considered as your vision will be obscured for too long a period. Instead, move over to the right-hand lane, where you have a much clearer view ahead, and complete your overtaking manoeuvre speedily before returning to the correct lane.

Facing page, bottom:
On this two-lane carriageway the view is so obscured that overtaking would be dangerous; bide your time until a better opportunity to overtake occurs.

Overtaking In Heavy Rain

Suppose you are travelling along a three-lane trunk road and find that you are behind a large vehicle in the left-hand lane which is throwing up a continual spray of water, making driving difficult. You wish to overtake but find that on pulling over to the middle lane your vision is still obscured by the amount of spray being thrown up to the off-side of the vehicle. The best solution is to move over to the right-hand lane where the visibility is unaffected, if it is safe to do so, and complete your manoeuvre before returning to the left hand lane.

If you are in a two-lane situation then it is prudent to stay behind the heavy vehicle because overtaking with obscured visibility is far too dangerous to contemplate. Bide your time until an opportunity does present itself for you to overtake in a safe manner.

DRIVING THROUGH FORDS

Should you come to a ford which is no deeper than the car's floor level you should be able to negotiate it quite safely. (Fords usually have a measured post indicating the depth of water.)

Reduce speed and select a low gear. With the engine running continuously to prevent the water entering the exhaust system, proceed at a slow steady speed so as to not create a bow-wave. If you go too fast and get water splashing up on to rotating parts it is likely that your ignition system will be short-circuited and will cut out the engine (this does not apply to diesel engined vehicles.)

Continue steadily until the water is safely negotiated and on reaching dry land operate your brakes several times to dry them out, otherwise you may find you have reduced braking power just when you may need it.

Flood Water

If you are unlucky enough to hit flood water or even a huge puddle at speed and your engine cuts out immediately it means that water has got on to the ignition system and short-circuited it. However providing your engine is thoroughly hot, ie: you have been travelling for some time, sit tight for a while as there is every likelihood that the heat of the engine will dry out the ignition system in about ten minutes or so and you will then be able to proceed.

It has been known for engines to be wrecked by water being sucked into the air intake and causing an hydraulic lock in the cylinders because of travelling too fast in such conditions.

As soon as you are clear of the flood water try your brakes two or three times to dry them out.

Right:
Cars and lightweight vehicles may use the bridge if the water is flowing too deeply, but others must use the ford. Note the depth post which shows that, despite the snow, very little water is flowing.

Below:
The correct way to negotiate a ford is by keeping the engine running and proceeding at a steady continuous speed without creating a bow-wave.

Below right:
After negotiating a ford there is often a reminder to try your brakes; apply them several times to dry them out.

DRIVING IN SOFT MUD OR SAND

If you have to drive over soft mud or sand the same driving technique is employed as for snow. Remember you may have to reduce the tyre pressures to increase traction and lessen the tendency to sink — and you must avoid wheelspin which will cause the vehicle to dig itself deeper into trouble.

If there is only one driving wheel spinning you will have to stop it because all the tractive effort will be going to this wheel via the differential. It may be that the spinning wheel is on softer ground and you will need to firm it up with whatever is to hand to get an even tractive effort from both driving wheels. Here again the drill is to select a low gear, be very gentle on the accelerator and drive the vehicle off the mud or sand at a steady, constant speed.

Don't let anyone stand behind the vehicle in case stones are thrown up from the spinning tyres. If you did reduce your tyre pressures remember to inflate them to their correct pressure.

Left:

The driver is unable to proceed because the off-side driving wheel is embedded in soft ground, creating wheelspin.

Packing stones around the wheel firms up the ground and prevents wheelspin.

DRIVING IN MOUNTAINOUS REGIONS

Be careful not to over-stress your vehicle when driving in hilly or mountainous regions. Select a gear which will enable the car to climb without labouring or stalling and on a long continuous uphill climb watch for signs of overheating. If at all possible pull in at a layby to give the vehicle a chance to cool. There is often a stretch of road allotted for this purpose and gives the driver an opportunity to view the scenery.

A Long Descent

When negotiating a long descent or mountain pass select a low gear and let the engine brake the vehicle. If, in spite of this, the vehicle gathers speed select an even lower gear. Excessive use of the brakes should be avoided as this can lead to overheating of the brake discs and drums, causing the brake fluid to vaporise with a subsequent loss of braking power known as 'brake fade'. Never

Above:
After enduring a long uphill climb on a very hot day the car is showing signs of overheating.

Left:
As this is a long descent the sensible driver will change to a lower gear to brake the vehicle and thus avoid using and overheating the footbrake.

Right:
An escape lane. These are placed strategically on long or steep descents so that drivers experiencing braking trouble have the opportunity to stop and avoid a catastrophe.

Right:
This is the wrong way to load cases as it gives a higher centre of gravity than is necessary and could contribute to the instability of the vehicle.

brake on a bend — always before – and don't get into a position of having to brake harshly, especially on a wet, muddy or loose shale type of surface as you will almost inevitably lose control of the vehicle.

Always take the opportunity of resting your engine when it has had to do some arduous work even if it only means driving at a cruising speed for a few minutes to let the engine cool down somewhat, as this is good for a vehicle in the long term.

DRIVING WITH A ROOF RACK

If you are planning to fix a roof rack on the car be sure to place the rack centrally to give equal weight distribution over the roof area and fit it securely to the individual roof rails or anchor points.

Place suitcases flat to keep the overall height as low as possible which keeps wind resistance to a minimum. Make sure that you

Left:

Travel with cases lying flat to give a lower centre-of-gravity. Place them in strong plastic bags with the closed end of the bag towards the front of the car and secure them with a spider or quick-release bungies.

place your suitcases or other equipment so that the load is evenly distributed and do not overload the rack.

Securing Your Cases

It is a good idea to cover your suitcases with a plastic sheet or put them in strong plastic bags to protect them, this is important during bad weather.

They can be secured with an elastic 'spider' or 'bungies', which are quick to secure and release.

Cross Winds

As many long journeys involve motorway driving which entails travelling speeds of up to 70 miles per hour, be aware of wind resistance which can be very strong, particularly if you happen to be driving into a high wind. Because your car is likely to be 'top heavy', cross-winds are another possible hazard which can affect the stability and steering of the car, so drive accordingly.

Remember to check the security of your suitcases and roof rack anchorages from time to time during the journey.

PARKING TECHNIQUES

The driving test now includes parking as well as turning in the road and reversing around corners.

Do not carry out any parking manoeuvres without first ensuring that it is safe to do so.

Parking in a Street.

A gap about 1½ times the length of your car should be sufficient space in which to park your car but it won't be enough if you attempt to drive forward into it. The rear part of your car is almost bound to protrude into the path of other road users, whereas the space can be sufficient for the car to be reversed into.

Traffic permitting, the correct technique is to drive slightly beyond the space with the car pointing to the middle of the road, then reverse into the centre of the space so that the left-hand rear wheel is a short distance from the kerb and the front of the car about to just miss the car ahead as it enters the space. Now put on a full steering lock which will bring the front over relatively quickly to the kerb as you continue slowly to reverse in, while at the same time the rear will come closer to the kerb.

It may be necessary to drive forward to give yourself an equal amount of space behind and in front of your vehicle in order not to trap the cars already parked.

Above:
The wrong approach. The driver attempts to drive forward into the kerb space.

Right:
The photograph shows that the space was inadequate for forward entry as the rear of the car is protruding into the line of traffic flow.

The right approach

Above left:
Step 1. Drive beyond the space and turn the wheels towards the middle of the road.

Above:
Step 2. Reverse into the space, applying full-lock so that the car's front just clears the other parked car.

Left:
Step 3. Complete the reversing operation, straighten the lock and leave space to the front and rear.

The wrong approach

Above:

Method 1 Step 1. The driver has seen a space at top left of the car park but the carriageway is not very wide.

Above right:

Step 2. He approaches from the right-hand side of the carriageway for as wide a left turn as possible.

Right:

Step 3. In spite of step 2 the driver cannot get the rear of his car in from this approach: the carriageway was not wide enough.

Parking In A Car Park

Because of the need to accommodate as many cars as economically as possible most car parks do not allow as much room to manoeuvre as most of us would wish.

Generally it is easier to reverse into a space. To get into a confined space you need to drive just beyond the space and reverse into it. If you can get the rear of the car in a space then the front will easily follow but it won't necessarily mean if you get the front in that the back of the car will follow.

Left:

Method 2 Step 1. Drive beyond the space — about a car's width - so that the car's rear is in line with the space.

Right:

Step 2. Quickly get a full-lock on as you reverse so that the rear is still in line with the space.

Below:

Step 3. Drive forward into the space.

Below right:

Step 4. See that you are centrally placed so that other drivers are able to open their doors.

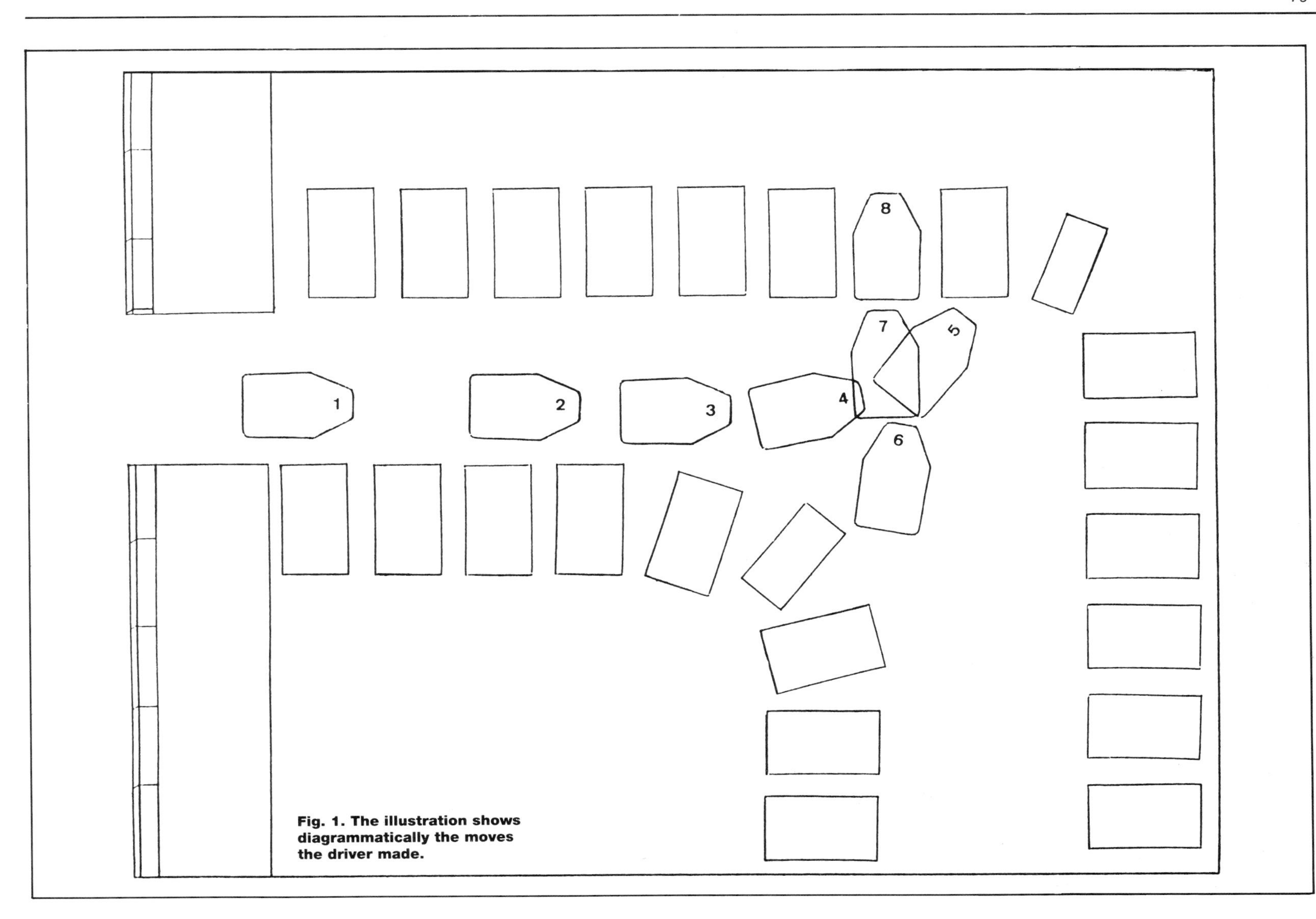

Fig. 1. The illustration shows diagrammatically the moves the driver made.

Above:
Method 3: Step 1. Drive up on the left of the carriageway, turning to the right just beyond the car space.

Above right:
Step 2. Reverse into the space.

Right:
Step 3. Centralize your car if necessary to allow other drivers access to their cars.

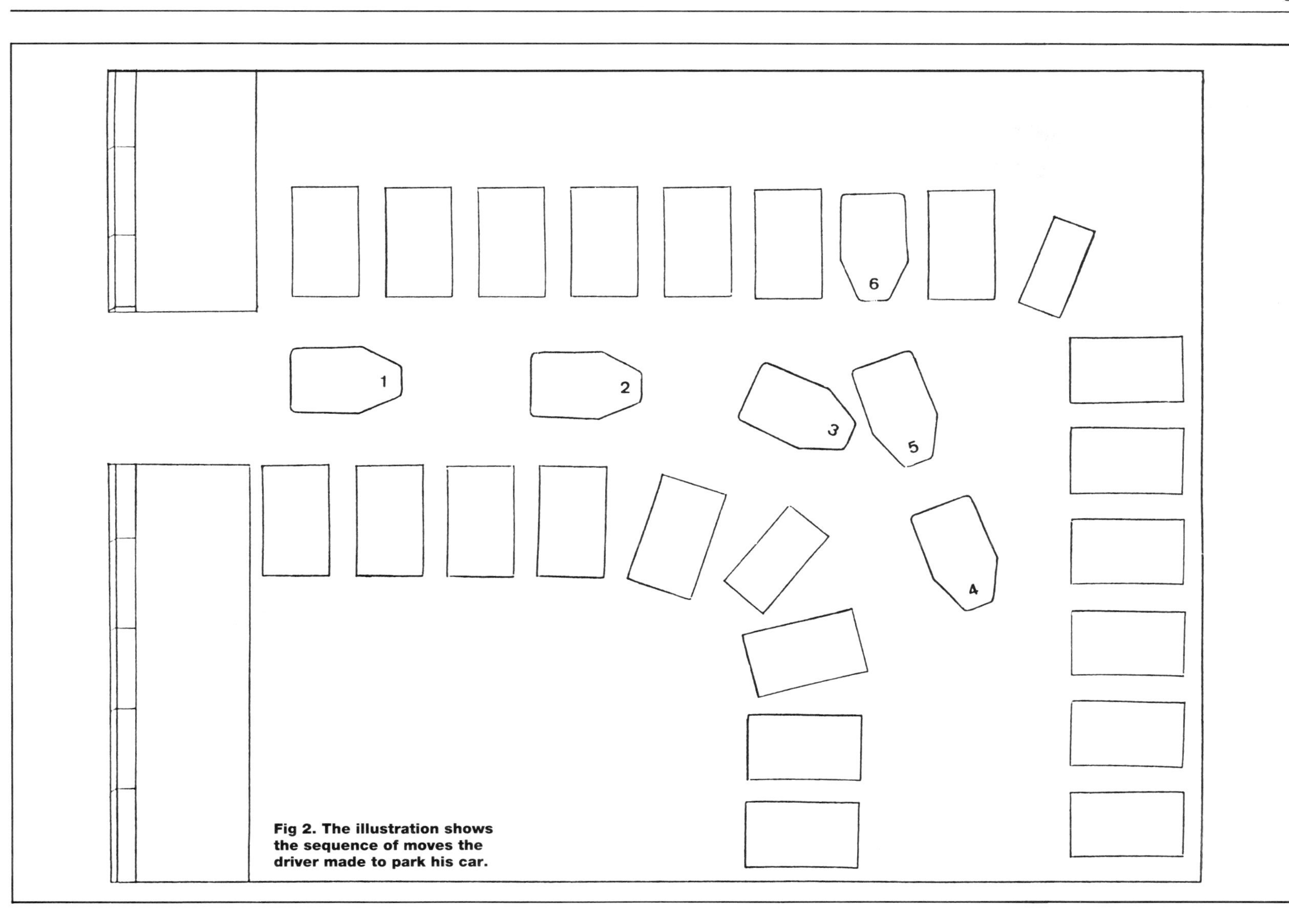

Fig 2. The illustration shows the sequence of moves the driver made to park his car.

Right:

Here is a good example of thoughtless parking. Not only is the car causing a restriction in the road, but it is also placed where pedestrians cross the road.

CONSIDERATE AND SENSIBLE PARKING

Despite the ever increasing amount of traffic on the roads a motorist can often help himself and others by considerate and sensible parking which benefits everyone in the long run.

For example, if cars are already parked along one side of a road used for two-way traffic and you wish to park, don't park opposite these vehicles and cause further restriction, even though there may be plenty of space and parking would be simple.

If at all possible, find a space on the same side as the already parked cars, which will still allow the passage of two-way traffic without restriction.

SUMMARY

Method 1, page 76. Should have been discounted by the driver upon entering the car park.

Method 2, pages 77-79. This manoeuvre was quite alright to execute, as this situation is encountered many times in car parks and other limited spaces.

Method 3, pages 80-81. This was quite the best method, Because the carriageway went to the right just opposite the car space the driver quickly assessed the situation and completed the operation in two (or maybe three) moves.

Left:

An unthinking driver has parked opposite the already parked cars although plenty of space is available with them. The car is now obstructing other vehicles, causing them to wait. This would not have happened if the driver had given a little thought to his parking.

6 Servicing

INTRODUCTION

Although the main servicing which covers all the major components of a car is usually done by the authorised dealer at the recommended mileage periods, there is also a need for regular minor servicing to ensure a reliable and economic running car.

The gauges and warning lights which monitor the behaviour of a vehicle should be watched frequently. For example, an engine temperature gauge showing a high reading may indicate that the cooling system needs to be topped-up, or the ignition warning light coming or staying *on* may mean that the alternator's driving belt is slipping.

Mind you, these things should not happen to a regularly serviced vehicle.

Whenever leaving or approaching your car check visually for signs of coolant, oil, fuel or other leaks.

Regular Maintenance

Charges for servicing and repairs to cars are high and these increase as the cost of the components rise, so a policy of regular checks should be adopted. Not only will you have a more reliable vehicle, but you will reduce running costs and may save yourself expensive breakdowns.

Routine maintenance to the engine should ensure economical fuel consumption and certainly a reduction of the pollutant gases. Tests have shown that 17% of poorly maintained vehicles produce 50% of pollution, and of these the worst 1% caused as much pollu-

Right:
A glance at the instruments shows that everything is functioning correctly.

Far right:
Upon returning to his car this driver has noticed that a leak has developed.

tion as 40% of the best maintained. New emission standards were introduced into the MOT vehicle test in 1991.

It is advisable that regular checks are carried out on the following items, where appropriate. It should be appreciated that not all the different models of cars can be covered here, but the following information will apply in some detail to almost any car.

SERVICING CHECKS

Checks are best done when the engine is cold.

As you get more familiar with your car and carry out regular checks you will have a better idea as to how often each component may need checking. Always investigate any new or unusual noises.

Above:
Examine pipes and cables for signs of chafing. Any faults spotted early can prevent a possible breakdown occurring later.

Engine Compartment

Open the bonnet and have a good look round. You do not need to be a motor engineer to observe that certain things are wrong, such as leaks from the radiator or engine block, hoses or cables chafing against other components, oil leaks developing, driving belts splitting or fraying, nuts or bolts working loose, etc. Have a general look and you may well notice a fault developing which could save a breakdown. Do keep the compartment clean; oil and dirt can cause problems, particularly with ignition components.

Engine Oil

Initially the engine oil should be checked every time you fill up with fuel at a service station but as you get familiar with your engine it may be sufficient to check your oil weekly. If the engine is hot, allow a few minutes for the oil to drain back into the sump but make sure the car is standing on level ground.

If a lot of oil is required you may have a bad leak which should be investigated and if blue smoke issues from the exhaust pipe it indicates a worn engine and that the car is burning oil.

Carburetter Hydraulic Damper (SU type)

On Austin Rover cars the carburetter hydraulic damper reservoir needs to be checked occasionally. Unscrew the piston damper assembly and add a small amount of engine oil (if necessary) to within about 12mm from the top of the tube. Then replace the damper assembly.

Hydraulic Brake Fluid Reservoir

The hydraulic brake fluid must be checked regularly although a sound brake circuit will rarely need topping-up. The fluid level will fall slightly over a long period and this will be due to the wear of the brake pads and shoes. A need to top up fairly frequently is a sure sign that a leak has developed in the brake circuit and must be investigated at once. *Do not let the reservoir get empty – you will have no foot brake.*

Hydraulic Clutch Fluid Reservoir

Check regularly, although very little fluid should be required over a very long period. If a significant topping-up is required, check for leaks in the system.

Facing page, top:

Check the engine oil dipstick for the correct oil level. When topping-up with oil, remember to replace the engine oil filler cap.

Facing page, bottom:

Check the SU carburetter hydraulic damper reservoir oil level. This is important as it ensures that the correct throttle opening responds to the pedal movement.

Left:

Look at the hydraulic brake fluid reservoir level. It should only need topping-up at long intervals but must not be omitted from regular checks.

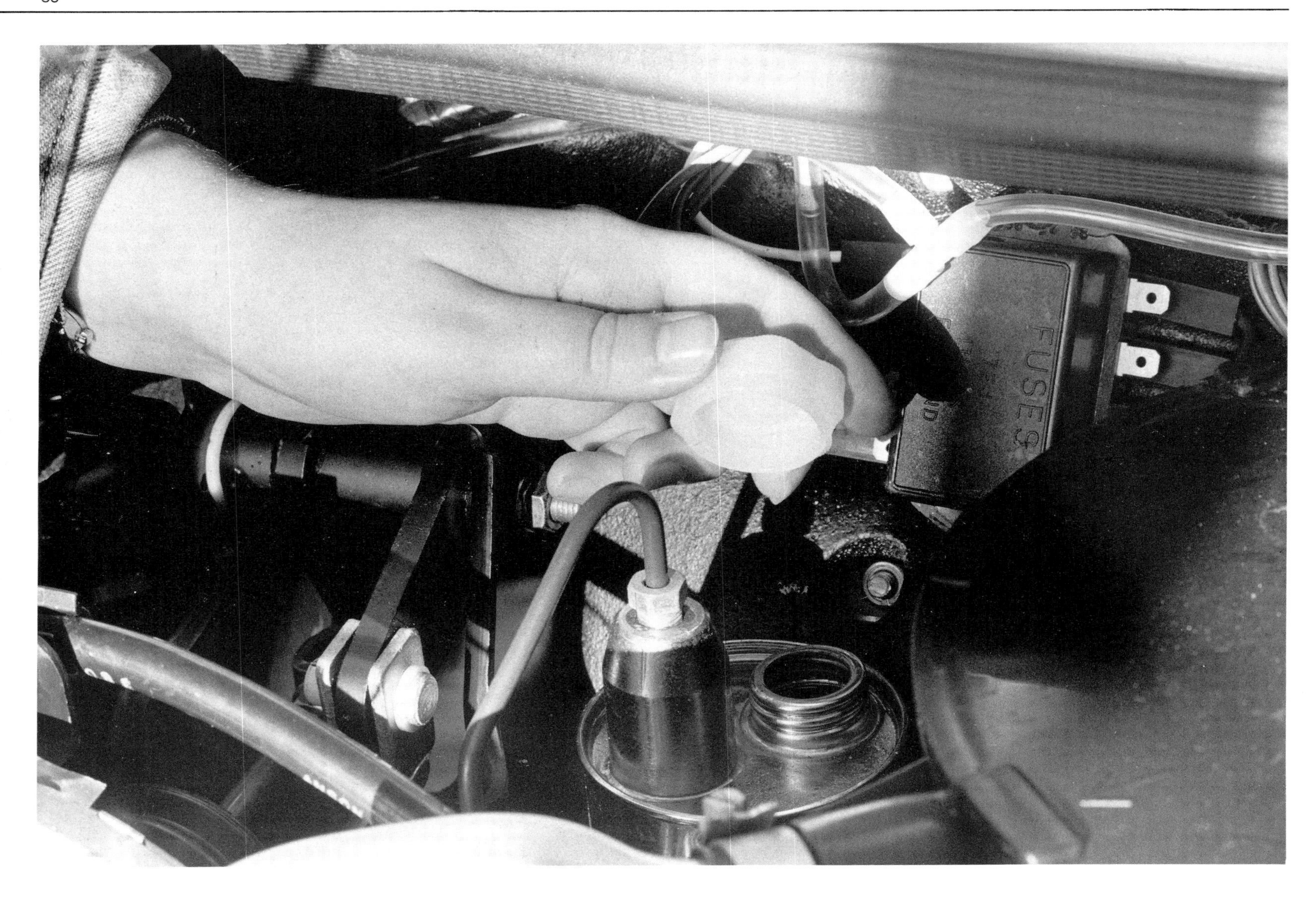
FUSES

Power Assisted Steering Reservoir

Check the fluid level and top-up to the 'max' mark if necessary. Do not overfill. It should rarely need fluid so frequent topping-up points to a leak in the system — investigate.

Radiator Coolant Level

Do not check when the engine is hot because the system is pressurised and you could be scalded. If you have acquired the car during the winter months make sure the system contains the correct solution of anti-freeze.

Check when the system is cold, turn the expansion reservoir filler cap a quarter of a turn to allow all the pressure to escape, then continue turning to remove the cap.

The expansion reservoir should be approximately half full, or there may be a level-mark indicated.

Should the system need topping-up fairly frequently, a leak in the system is indicated. When replacing the cap it is important that it is tightened down fully, not just to the first stop.

Windscreen Washer Reservoir

Remove the reservoir cap and top-up as required, leaving an air space — say 25mm — below the filler neck. It is helpful to add a windscreen washer solvent which assists in removing mud, flies and road film. In cold weather methylated spirits or any proprietary aid can be added to prevent freezing of the water. *Do not allow the reservoir to get empty — otherwise it may be difficult to prime the system.*

Battery Check

Many modern cars are now fitted with a maintenance-free battery with a sealed cell top-well. Although these should not need any maintenance, make sure corrosion is not building up on the battery connections. Batteries with removable filler plugs may need topping-up from time to time. Add distilled water to a level about 10mm above the cell plates. Batteries tend to want checking more frequently in hot weather and as they get older. If checking in the

Far left:
Examine the hydraulic clutch fluid reservoir level. It is not likely to need topping-up very often.

Below left:
Check the power-assisted steering reservoir for the correct oil level.

Below:
Check, when cold, the expansion reservoir for the correct coolant level. Ensure the cap is replaced and returned to its final stop.

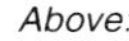

Above:

Top-up the windscreen washer reservoir as necessary. Other constituents may be added to clean off road filth and to offset freezing of the water in winter.

Above right:

Add a little distilled water to the battery cells. Top-up to a level of about 10mm above the plates.

dark or subdued lighting, avoid using a naked flame as batteries emit hydrogen.

Make sure battery connections are corrosion free; clean with soapy water if necessary and smear the connections with vaseline.

Some maintenance-free batteries may still have removable filler plugs and is a good idea to check the cells occasionally. It sometimes happens that the electrolyte is below the top level of the plates and a longer life is ensured from the battery by topping-up with distilled water very occasionally.

Tyre Checks

Check the tyre pressures regularly when the tyre is cold and inflate to the manufacturer's recommended pressures. Don't forget the spare wheel. Premature wear and possibly a blow-out will occur if you drive on under-inflated tyres.

Check for cuts and damage to the side walls. Check that your tyres have a full width tread and have not worn beyond the legal requirements.

If carrying a load on the roof rack or setting off on a fast drive on a motorway, increase the tyre pressures by about 0.1 Bar or to the pressures recommended by the manufacturer. Deflate to regular pressures when returning to ordinary motoring.

Lights, Windscreen Wipers and Horn

As all these components are in constant use a specific check is not really necessary. Check, however that wiper blades are not becoming worn, and ask someone to help you check your lights — especially indicator and brake lights.

Check also to see that all lenses and glasses are clean — a substantial amount of light can be lost through muddy headlamp glasses.

Locks, Hinges and Catches

Lubricate locks occasionally by putting a spot of oil on the key, inserting the key and turning it — do this several times. Smear a tiny amount of grease on the working surfaces of the door catches — but not enough to get on your clothes when entering or leaving. Lubricate all hinges, bonnet and tailgate catches.

Seat Belts

Frequent inspections should be carried out to ensure that the effectiveness of the belts is not impaired. Inspect the belt webbing periodically for fraying or wear and check that the fixing points are secure. Check also that the buckle and tongue locking function is correct, and the buckle stalk cabling anchorage is secure.

Left:

Check tyre pressures all round. Look out for side wall damage, cuts or blisters, or exposure of the ply or cord structure. Remember, under-inflated tyres increases fuel consumption.

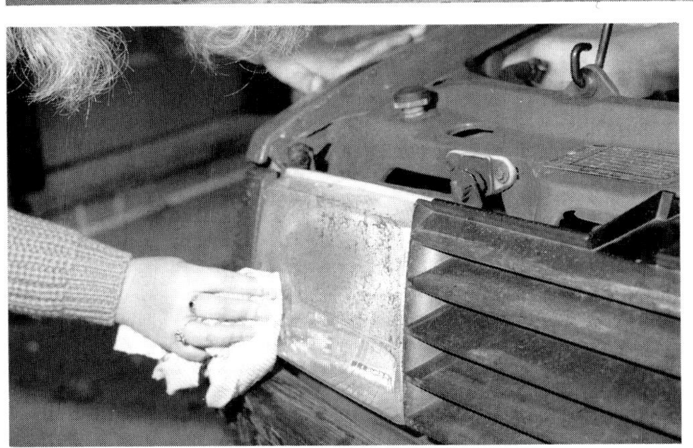

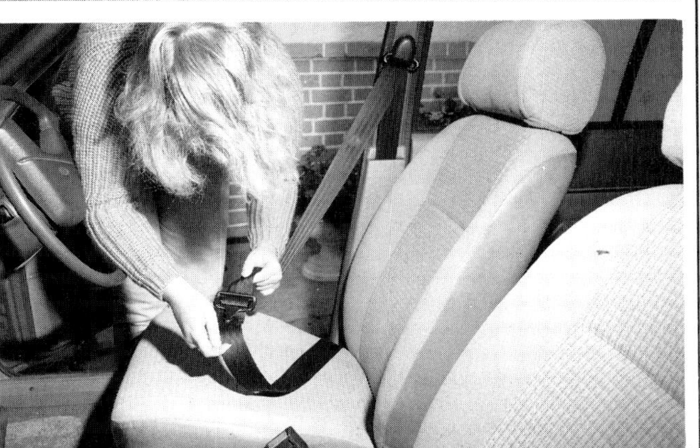

Far left:

Clean road filth from headlamps. Up to 50% of light can be lost owing to dirt on the glass.

Left:

Examine seat belts for fraying and the anchor points for security. Grubby belts will soil passengers' clothes.

Right:
Top-up with anti-freeze to ensure the correct mixture strength for low winter temperatures.

Centre right:
In very low temperatures you can leave the handbrake *off* overnight to prevent it freezing in the *on* position. *But do chock the wheels.*

Far right:
Ease off the wheel nuts with added leverage, after having had a new tyre fitted. Retighten the wheel nuts yourself to ensure that you will be able to loosen them in future.

CARE OF THE CAR IN WINTER

Many problems arise during winter solely because of the cold damp weather. Taking the following precautions will help to eliminate most seasonal faults.

1. Ensure that the correct grade and adequate amount of anti-freeze is mixed with the coolant to resist quite low temperatures. Remember when travelling along the highway that the slip-stream temperature is much lower than the ambient temperature, which is why one sometimes sees cars at the roadside with their radiators boiling even on a cold day. (The bottom of the radiator has frozen and is not allowing the coolant to circulate).
2. Use a light viscosity oil in winter — say 10W/30 multigrade engine oil. The engine cranking speed will increase because there is less drag when operating the starter. This makes starting easier and relieves the battery of some of the heavy load imposed on it.
3. Keep the battery fully charged, especially in diesel engine cars.
4. Adjust your air cleaner to its winter setting (if appropriate to your model).
5. Keep plugs and points in good condition.
6. Use a damp-removal aerosol or a spray which coats the ignition leads with a protective lacquer.
7. A newspaper or a sheet of plastic when placed against the wind-screen will prevent it icing up when parked for some hours.
8. A fine film of petroleum jelly smeared over headlights will prevent snow collecting on the glass.

9. Petroleum jelly may also be used in the door locks to prevent freezing and allow the key to be inserted.
10. If parking for the night on level ground on your own property leave the handbrake *off*, otherwise it can freeze and be difficult to release, *but do chock the wheels*.
11. Finally, if you live in a region where fairly long severe winters are usual change to winter grade tyres at the onset of bad weather, or fit all-year-round tyres permanently.

TIGHT WHEEL NUTS

A common complaint from motorists who have had new tyres fitted by tyre agents is that they have been unable to loosen the wheel nuts subsequently when sustaining a roadside puncture.

To obviate this problem it is a good idea to loosen the wheel nuts when you get home. Retighten them securely with your car's wheel-nut spanner so you'll know that you will be capable of loosening the nuts in future.

CORRECT USE OF DIFFERENT TYRES

The use of cross-ply or radial-ply tyres can be safely fitted on all cars in the following sequences.
1. Cross-ply on all four wheels is safe.
2. Radial-ply on all four wheels is safe.
3. Radial-ply on rear wheels with cross-ply on front wheels is a safe mixture, but is not suitable for high performance cars.
4. *Never* put radial-ply on front wheels with cross-ply on rear wheels.
5. *Never* mix tyres on the same axle.
6. Different tread patterns make little or no difference to performance but it is sensible and desirable to try and keep to the same tread pattern.

SUMMARY

Most of the foregoing checks take only a few minutes and are very easy. Now that a lot of reservoirs are made of transparent plastic and the fluid levels can be seen at a glance a visual ckeck will suffice — just bounce the front of the car to see the fluids slopping around.

Regular maintenance pays, and do use only the manufacturer's recommended lubricants and fluids.

Just because a component hasn't needed a top-up for a long time, it does not mean that it can be overlooked; by checking you ascertain that it is still functioning correctly.

7 Car Valeting

Right:

Wash down bodywork with a sponge and copious amounts of water; the brush is for any awkward spots or crevices. Note any paint blemishes for rectification later.

INTRODUCTION

Keeping a car in a clean, serviceable condition tends to be a constant battle against the elements, but one of the good things about washing your car regularly is that it gives you an opportunity to monitor the condition of the bodywork and take the appropriate action regarding any paint blemishes, or arrest any corrosion before it gets too bad.

BODYWORK

Wash the bodywork frequently, using a sponge with copious amounts of cold or lukewarm water. Never use household soap or detergents although you will find that any approved car shampoo will be helpful in removing traffic film. After shampooing, rinse with cold water and dry with a clean chamois leather.

Tar spots can best be removed by gently rubbing with a cloth dipped in a mixture of three-quarters petrol and one quarter clean engine oil. Wash off with clean water and dry with a clean chamois leather.

If you use a hose, avoid directing it at full force against the body.

A vinyl-covered roof can be washed with tepid water and toilet soap, using a soft brush to remove dirt from the grain finish. Do not use other cleaning agents.

An occasional polish with a quality wax car polish will give the paintwork extra protection. For maximum protection, polish the bodywork at least twice in the autumn, before the onset of winter.

Any bright metal or chrome parts are best cleaned with a proprietary chrome cleaner.

Winter Wash

Winter time poses a problem as roads are often swimming in slush and salt, and a car can become filthy on one journey. Don't let this

Below left:
Remove tar spots with a soft rag dipped in a made-up cleaning agent.

Below:
Waxing a car regularly gives a protective coating and gives lasting benefit to the bodywork.

Above:
Don't allow winter dirt remain on the car long; frequent hosing down with cold water should suffice.

Above right:
Finish off with the brush-head removed and concentrate the jet under the wheel arches to remove snow and road salt.

Far right:
Garage vacuum cleaners are more powerful than domestic cleaners and are ideal for removing dog hairs or obstinate particles.

condition last too long and if possible wash down the bodywork frequently with a car hose, using the brush to loosen the filth and flow of the cold water to wash the filth away. This takes only a few minutes if done regularly.

Remove the brush head and play the hose jet underneath the bodywork, concentrating on the wheel arches and any little pockets in the underbody parts. It is in these vital areas that body corrosion is likely to start if dirt is allowed to remain.

At the end of a winter it is a good idea to have the whole of the underneath steam cleaned.

Interior

The easiest way to keep the interior of a car clean is to vacuum it regularly. A receptacle to keep toffee papers, apple cores etc. in is a great help.

Cigarette stubs which are still alight should never be thrown out of the window as they could cause a fire by coming into contact with petrol which has leaked on to the road.

Stains can be removed from the upholstery, whether it be PVC or leather, by the use of one of the many upholstery cleaners on the market, but never use petroleum or a spirit solution on any interior trim.

Endeavour to keep the seat belts clean as they could mark your clothes as they chafe during a journey. The best way to clean a belt is to sponge it with warm water using a non-detergent soap and allow it to dry naturally. Do not use caustic soap, chemical cleaners or detergents, and do not dry with artificial heat or by direct sunlight.

Vacuuming

Many domestic vacuum cleaners are not powerful enough to suck up those annoying pieces of fluff which seem to be glued to car carpets and it is a good idea to occasionally use the commercial vacuum cleaner now available on most garage forecourts for a small charge.

These are powerful machines and they make a good job of cleaning the carpets and upholstery.

Windscreen/Windows

Hinge the wiper blades away so that you have a clear screen to work on. Don't use washing and polishing agents which contain silicone, as this tends to cause smears on the glass which reduce visibility, particularly during darkness and in rain.

VACUUM
LASER JET

Right:
Remove frost with a de-icer spray, but if this is not available a plastic scraper will suffice. Never use a metal scraper.

Right:
Wash road dirt and insects from the windscreen with soapy water before spraying on a proprietary cleaning agent specially produced for windscreens.

The rubber blades can be damaged by contact with material such as cleaning agents, grease, silicones or fuel.

Clean all glass surfaces regularly by softening the traffic film and road dirt with soapy water to prevent scratches and then apply vigorously a proprietary brand of cleaning agent with a damp sponge. Finally, rinse with clean water or wipe off with a soft clean cloth.

Wipe the blades to remove any imbedded grit before lowering them to the screen.

Heated Rear Window

Clean the inside of the rear window with a soft cloth or a damp chamois leather to avoid damaging the heating elements. Never use solvents or sharp objects to clean the glass. Remove frost or ice with a de-icer spray, or with a plastic scraper; never use a metal scraper.

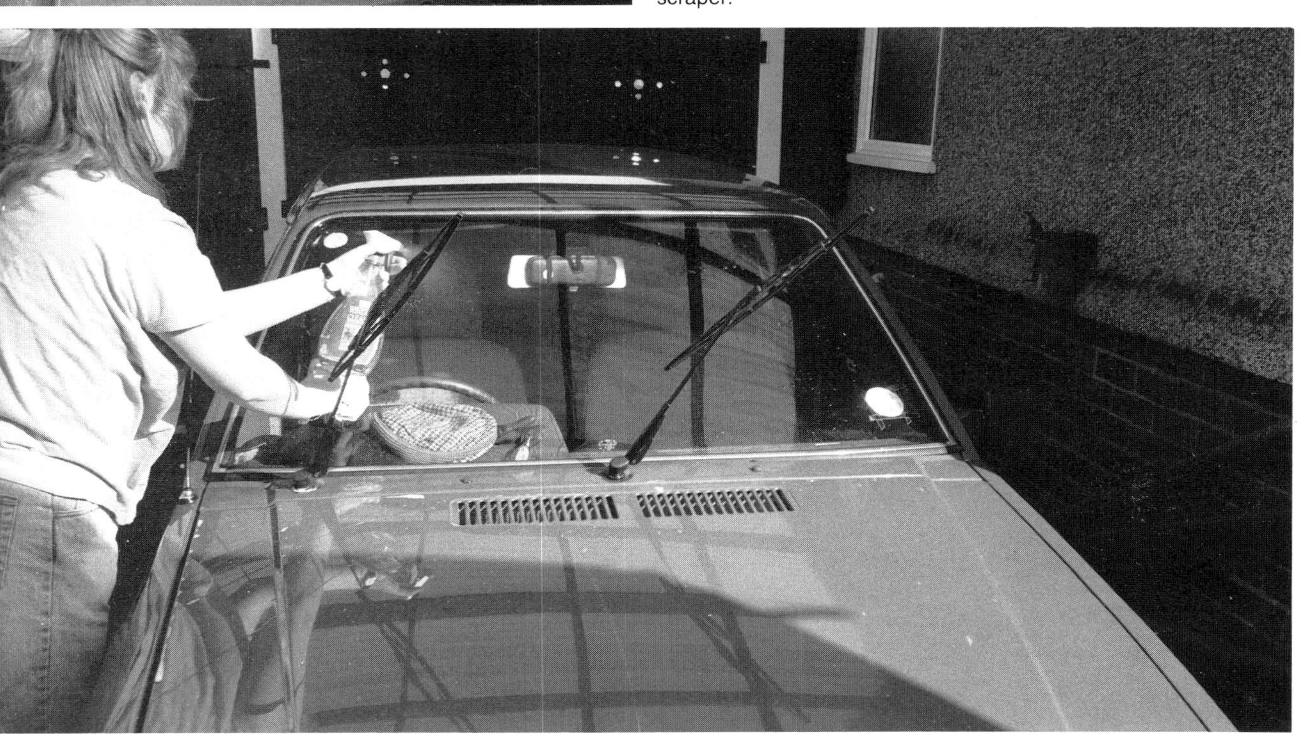

Left:

Treat minor paint damage caused by flying stones with a painting stick. Aerosols are suitable for spraying large areas requiring more than just a touch-in.

External Rear View Mirrors

Care should be taken to prevent scratches to exterior rear view mirrors.

Always soften dirt and mud with soapy water before washing mirror surfaces and polish with a soft nap-free cloth.

Never use abrasive cleaning compounds on mirror surfaces.

Plastic Surfaces

If plastic surfaces do not appear to be clean after a normal wash special cleaning preparations are available. Apply the agent according to instructions and brush the surface intensively with a hard brush. Rinse off with warm water and dry with a chamois leather.

When polishing the vehicle make sure that no polish comes into contact with the plastic surfaces as it will be difficult to remove.

Paint Chip Damage

The paintwork should be inspected periodically for chips or scratches and is best done when the car has just been cleaned. Pay particular attention to the front and sides of the car which may have been chipped by stones thrown up by your own or other vehicles.

The edges of the doors may have been chipped or scratched if opened against a wall or other obstruction.

Small paintwork blemishes or rust spots should be dealt with straight away to stop further deterioration.

Above:
Aerosoles are suitable for spraying large areas requiring more than just a touch.

Painting Sticks

Bodywork paint for small areas can be bought in painting sticks from any good motorist's shop stocking a comprehensive range of the most popular makes, or from the spares department of any main agent dealing in your make and model.

Nowadays these paints also have a rust inhibitor included, so they can be applied directly to minor blemishes.

These painting sticks have a conveniently designed screw-cap with an integral touch-in brush.

The sticks must be well shaken to get the correct shade of colour. If you fail to shake long enough you may find that the paint has not mixed to its proper shade. For larger areas of damaged paintwork aerosols can be purchased. These give instructions on how to prepare the surfaces to be sprayed, how many coats to use and at what intervals.

CAR WASHES

If you do not want to wash your car manually there is always the option of an automatic car wash. There are an increasing number of these about now and they have much improved since their early days, when they had a reputation for scratching paintwork and tearing out aerials or wiper blades. Even so, the best car washes still leave some nooks and crannies untouched, inside the doors for example, and these you will have to clean yourself.

A notice prominently displayed near the wash should tell you the procedure to follow. You do need to remember to close all doors properly, shut windows and sun-roof and check that the wipers are stowed, in the normal horizontal position. The aerial should be retracted and the roof rack, if fitted, checked for security, before driving through the wash.

If using the wash for the first time it is a good idea to enquire at the Service Station whether there is any particular action to take; such as, do the wipers need to be stowed horizontally or vertically, as this will depend on the design of the washing apparatus.

The most common types of car wash available are the roll-over type and the hand-held lance type.

Most modern car washes are now designed to filter and recycle the water to enable it to continue to operate during a drought.

Roll-Over Car Washes

Different programmes of wash can be selected with the roll-over version. The machines are capable of washing, drying, applying shampoo foam, waxing, hot waxing, wheel brushing, under-chassis washing and under-chassis waxing. Programmes consist of various combinations of these and the programme combining most tasks will naturally be the more expensive.

The most expensive wash first covers the car with shampoo foam, which is quite effective in loosening any diesel film or road filth, before it is washed off. The wheels are cleaned with contra-rotating brushes and cleaning of the under-chassis takes place.

Far left:

Prepare the car by ensuring doors, windows, etc are closed, and that wipers are stowed. On this model the aerial can be unscrewed.

Left:

Insert the token in the machine before correctly positioning the car in the wash.

Right:

The revolving brushes travel over the full length of the car and return; jets of water remove dirt.

Next, the car is covered with a super wax which achieves both a shine and forms a protective film. The whole car is then hot air-dried.

Even the cheaper wash without shampoo does an adequate job as the water usually contains some chemicals and detergents that shift the grime more effectively than plain tap water.

HAND-HELD LANCE CAR WASHES

There are various types of hand-held lances. One early type has a control in the handle that enables one to change the force of the jet or its spray pattern. This is particularly suitable during the winter, and particularly if the vehicle is subject to a lot of motorway mileage.

It is sensible to wear wellingtons when using the lance. Use the gentler spray for the bodywork, thoroughly soaking and loosening the road grime before removing it. This may take longer than the powerful jet, but the force of the water playing on any road grit is likely to damage the paintwork. The jet gets behind the road filth or grit, forces it away and in so doing spreads it, causing minute scratches to the paintwork.

The high-pressure jet is most useful for cleaning the underbody, particularly in harsh wintry conditions. Make a point of using it frequently and washing thoroughly under the wheel arches and the surrounding areas, especially if salt has been put down to combat snow and icy road conditions.

The latest and more sophisticated type of lance offers a number of automatic programmes, such as shampoo and rinse, hi-foam and rinse, shampoo and hi-foam and rinse, and shampoo and hi-foam and rinse and wax and hot shampoo and hot hi-foam and rinse wax.

The machine has two hoses. One is a lance hose with an *on/off* control in the handle and the other is a brush and shampoo hose which is controlled by a timer within the machine.

Payment for the programme you intend to use is made at the cash desk for which you will receive a token; the more comprehensive wash is more expensive than the simpler one.

The automatic wash starts when you put your token in the slot. Symbols displayed on the machine explain the sequence of operations for your particular programme. From then on you do as the machine dictates.

Adequate time is allowed for cleaning the car and it is recommended that you begin from the lower edges and work upwards.

Immediately after the wash remember to apply your foot brake several times to dry out the brake shoes.

8 Accidents – the action to take

Below:
For a one-year-old baby a child's seat secured by the adult seat belt, with the baby facing rearwards, is considered the safest method to adopt.

INTRODUCTION

Sad to say, about 5,000 people are killed on our roads each year, many of them innocent victims of another person's selfish and arrogant driving manner. There are also 54,000 serious and 180,000 minor accidents each year and most are avoidable.

Governments have been wrestling with the problem of reducing car accidents for years and manufacturers have been encouraged to improve the safety of their cars; much research is being carried out to this end.

Seat Belts

To date, the biggest single contributory factor in the reduction of fatal accident numbers has been the compulsory wearing of seat belts, introduced by legislation in 1983. Initially only front seat passengers had to wear seat belts. However, many manufacturers have been fitting rear seat belts since 1987; but not until 1 July 1991 when further legislation was introduced were rear seat passengers compelled to wear seat belts. This means that all passengers travelling in cars with the registered mark 'E', and all subsequent registration letters, must wear seat belts, as well as those travelling in cars whose owners voluntarily had rear seat belts fitted before 1987.

Responsibility for wearing seat belts rests with any passenger over the age of 14 years and failure to do so can result in a fine of £100 max.

There is no retrospective action for cars not yet fitted with restraints.

Under certain circumstances there is sometimes a drawback to wearing a seat belt as the belt itself can cause injury, but there is no doubt that the advantages of wearing a belt far outweigh the disadvantages.

Children's Restraints

Since 1st September 1989 it has been the *driver's* responsibility (*not* the parents if neither parent is driving) to see that children

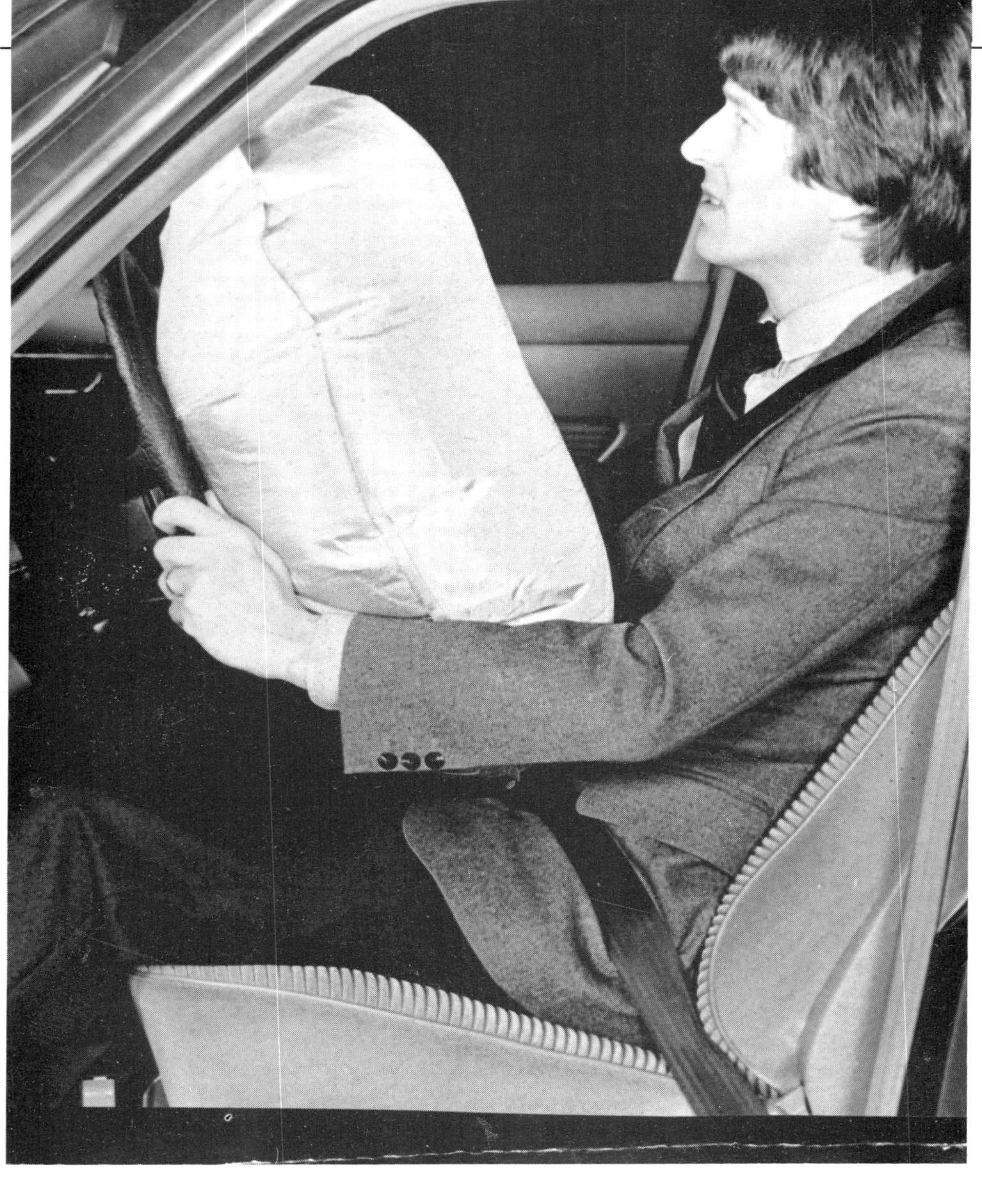

under 14 years of age use restraints if fitted in the rear of the car. Babies under one year old may travel in a carrycot or an infant carrier which is restrained by straps.

Children of one, two or three years of age may travel in an appropriate child's seat or harness, or on a booster cushion used in conjunction with an adult belt. An adult's belt perhaps with a cushion, is suitable for children from four to 14 years of age if suitably adjusted.

Ideally, a child should be restrained in a purpose designed restraint appropriate to the child's weight. The restraint should be labelled to show the weight for which it has been designed.

Important! If there is a child restraint in your car and there is space for it to be fitted without the use of tools, it is considered to be available for use by a child of the appropriate age and weight. IT MUST BE USED. If it is not used, you risk a £50 fine.

Airbag

Another concept for minimising injuries to drivers when a head-on collision occurs is the 'airbag'. An airbag is fitted into the centre of the steering wheel and is designed to inflate within 1/27 of a second upon impact to cushion the driver from injuries to the face and chest.

The system is designed to react only if the sensor registers an impact of a strength corresponding to a severe frontal collision. It is not activated by minor impacts or damage.

The system acts extremely fast. Inflation and deflation of the airbag is completed in a fraction of a second and the driver is in contact with the airbag for a moment only.

Air bags designed to protect the front seat passenger are now also available and in America 2½ million cars are already fitted with these bags.

FAILING TO REPORT AN ACCIDENT

It is sometimes possible to avoid being convicted for failing to stop or failing to report an accident if you can convince the court that you were not aware that you had been in a collision. For example if you were driving a large, heavy vehicle and your tail-end scraped another car in the next lane it could be claimed that, because of the size of the vehicle, you were unaware of any incident. However, this line of defence is unlikely to succeed if you were driving a car.

Failing To Stop — Penalties

The maximum fine for failing to stop after an accident is £2,000. Upon conviction, you could expect a fine of at least £125 and, depending on the seriousness of the circumstances, the penalty points endorsed on your licence will be between five and nine.

If it is suspected that alcohol was the main reason for failing to stop, a disqualification period may be imposed.

Failing to Report — Penalties

The maximum fine for failing to report an accident is also £2,000 and upon conviction you can expect a fine of at least £125. The range of points which can be imposed, according to the seriousness of the offence, is between four and nine. Where both offences, failing to stop and failing to report, have been committed the Court will in all probability consider a period of disqualification.

Accidents Involving Animals

If you are involved in a collision with an animal, whether it be a horse, sheep, cow, pig or dog — but not a cat — you must stop and if possible exchange details with the owner. If you are not able to locate the owner you must report the incident to the police.

ACCIDENT PROCEDURE

If you are involved in an accident certain steps should be taken to ensure that your subsequent accident claim goes through without a hitch. First of all always stop and exchange names and addresses. (It's a good idea to carry a pencil and pad in your glove compartment at all times.)

If no personal injuries are sustained, in theory there is no need to call the police. However, complications can arise either if the other driver involved admits the blame for the accident at the time then later changes his mind or doesn't want to exchange details and drives off, or says that no damage has been done so it doesn't matter. It is best, therefore, to try and find an independent witness or if you have any doubts about the other driver, report the incident to the police within 24 hours to safeguard yourself.

Always bear in mind that if the other driver can take you to Court on a charge either of failing to stop/report or, worse still, driving without due care and attention, he is more likely to be awarded damages through a Court or be able to convince his insurance company that he was not to blame.

Obviously, if you are badly injured you will not be able to follow the following advice. However, as many accidents only result in structural damage to the vehicles concerned, we will assume that, having been involved in an accident, you are shaken but able after a minute or two to take down some necessary particulars. The procedure is as follows:

1. Exchange drivers' names and addresses. Although you are not obliged to give the other driver your insurance details, it is wise to

Far left:

The photograph shows how the inflated airbag cushions the driver and prevents him from injuring himself on the car's steering and bodywork.
Courtesy Mercedes-Benz

Left:

Notices appealing for witnesses often appear at the roadside when police are investigating a serious accident.

Right:
A car which cannot be towed after being damaged in an accident is usually recovered and transported home by this method.

do so since it may discourage the other driver from calling the police.

2. Try to find independent witnesses and take their names and addresses. These should be passed to the other driver if he hasn't already recorded them.

3. Note:

The position on the road of both cars — middle, near kerb etc — and the time of day.

Road conditions, any obstructions, road works, road signs, etc.

The speed at which you and the other car were travelling and the weather conditions.

Note any parked vehicles or circumstances which might have contributed to the accident – such as a child on a bicycle, a pedestrian in the road, etc.

4. Make a note of the other car's make, model, colour, registration number, extent of damage and, in particular, the condition of his

tyres. If you happen to have access to a camera, take a photograph; it could prove invaluable.

5. Finally, report the accident to the police to safeguard yourself. This should forestall the driver who refuses to report the accident to his insurance company and makes no subsequent claim. If he should be uninsured the police will prosecute him and this will be to your advantage if you should have to seek re-imbursement through a court.

Left:
In this incident the two drivers were too badly injured to be able to exchange details and all particulars were taken by the authorities.

INJURIES TO PEOPLE

If, while driving, you are involved in a collision in which someone sustains injury, you are obliged to report the accident to the police as well as exchanging names and addresses and all the relevant particulars. The details must be reported by the driver, not by someone else, to the police at the scene or at a police station as soon as possible.

If the police are called to the scene it is likely that they will ask you to explain what happened and to make a statement. You should remember that anything said to a police officer is likely to be recorded by him in his notebook either at the time it is said or subsequently. Any statement made may be referred to later. So you should think carefully before saying anything. If you are pressed for a statement, it might be wise to write it out yourself so you can be sure of what you want to say.

Above:

Fortunately mainly structural damage was sustained in this collision and particulars were exchanged by both drivers, as is required.

Right:

Any damage done to someone else's property must be reported to the police.

MOTORISTS LEGAL PROTECTION LIMITED

62-72 Victoria Street, St. Albans, Herts. AL1 3XH. Tel: (0727) 69152. Fax: (0727) 61206

PROPOSAL FOR 'AFTER THE ACCIDENT' POLICY

pleting this proposal your are under a legal duty to disclose all material information — that is, information which is likely nce the acceptance of your proposal and/or the terms applied. Should you fail in your duty, the insurance could be invalidated. have any doubts as to whether any item of information is material, then you should make full disclosure of it.

uld keep a record (including copies of letters) of any information you provide. A copy of your proposal form will be provided uest.

men policy is available on request.

Please complete all questions in BLOCK CAPITALS

OPOSER

Name:
Address:
Occupation: **Age:**
Phone: (Day) **(Evg)**

BROKER STAMP

Person dealing **File Ref**

TAILS OF ACCIDENT

Date: (b) Time: (am/pm):
ce (precise):

IMPORTANT: A COPY OF YOUR MOTOR INSURER'S ACCIDENT REPORT FORM MUST BE ATTACHED

TAILS OF YOUR MOTOR VEHICLE

Reg. No.: (b) Make, Model, Year:

UR MOTOR INSURANCE DETAILS

Name of Insurer:
Address:
Policy or Cover Note Number:
Cover: (Tick) Comp: ☐ Write here any excess £ Third Party, Fire and Theft: ☐ Third Party only: ☐
Is your No Claim Discount protected or guaranteed? (YES/NO)

UR OWN LOSSES OR DAMAGE Please estimate **your own** losses

(a) Vehicle Total Loss (ignore if Comp cover)	(b) Vehicle Repairs (ignore if Comp cover) obtain 2 garage estimates	(c) Policy XS (Enter only if Comp Cover)	(d) Damage to Clothes Tools, etc.	(e) Total estimated Loss of Earnings

Is your vehicle legally and safely drivable? (YES/NO)
Will you NEED to hire a replacement vehicle? (i) Before repair? (YES/NO) (ii) During repair? (YES/NO)
EASE NOTE: You will have to bear the cost of any hire yourself initially. Justifiable hire charges can then be included in your im against the third party.

UR OWN INJURIES

EASE READ CAREFULLY THE PART OF THE LEAFLET DEALING WITH INJURIES
scribe any injuries you have suffered:

OSSES AND INJURIES SUFFERED BY OTHER PERSONS WHOM YOU WISH TO BE HELPED UNDER THE POLICY

escribe any injuries or loss of or damage to property suffered by any other person/s in or on your vehicle in this accident:

Name	Mr/Mrs/Miss	Age	Injury & type and amount of property loss or damage
1			
2			
3			
4			

OMMUNICATIONS WITH THE THIRD PARTY ETC.

ave you any reason to suspect that the third party is not insured or that he or his insurer is not able or prepared to settle any aim against him? (YES/NO)
YES please state your reasons:-

as there been any correspondence with the third party or the third party insurer concerning your claim or the accident?
ES/NO) If YES, all such correspondence **must be attached.**
as there been a denial of responsibility for the accident from the third party or his insurer? (YES/NO)
YES please give details:-

However, if a policeman is not at the scene, you must report the accident straight away and will be required to produce your driving licence, your certificate of insurance and a current MOT certificate (if applicable) at the police station within seven days. It then rests with the police whether or not further action will be taken.

Left:
This shows the type of proposal form you will receive for an 'After The Accident' policy.

DAMAGE TO PROPERTY

If you have damaged a parked car or someone else's property you must either endeavour to find the owner to give him details of your insurance, or stop and make sure someone knows you have tried to find the owner and take his name and address. If you cannot do either of these things for any reason leave details of your name and address at the scene.

Report the incident to the police as soon as possible, certainly within 24 hours.

MAKING A CLAIM

Having obtained all the necessary details concerning an accident you will need to inform your insurers who will then send you an accident claims form.

The form requires you to disclose all details of the accident and you will be required to obtain an estimate of repairs from two different sources.

When your insurers have received these they will authorise the repairs, assuming that your policy cover is Comprehensive.

If it is imperative for you to have the use of another car whilst your car is in for repair then your insurers may give you permission to hire a car, for which they meet the expense within your claim.

If you only have Third Party Cover or Third Party Fire and Theft Cover you should still inform your insurers of the accident, but you will be responsible for your own repairs and claiming back the costs from the negligent party.

After-The-Accident Insurance Policy

There are policies now available which can be taken out *after* an accident. The object of such a policy is to enable you to claim for losses or expenses incurred in pursuing your original claim.

Even if you are comprehensively insured there are likely to be expenses incurred which are not covered by your policy.

If you are only insured for Third Party Fire and Theft it may be worthwhile taking out one of these policies, particularly if personal injuries have been sustained (see Chapter 3).

9 Breakdowns – some tips

Right:
The car has broken down and the driver has placed a warning triangle some distance back to warn other drivers that something is amiss.

INTRODUCTION

Good maintenance reduces the likelihood of a car breakdown, but most drivers will experience this sometime or other. However, certain precautions can be taken to mitigate the circumstances; one is to carry adequate spares and tools to tackle a minor breakdown.

Always protect yourself as much as possible by placing triangles, cones or flashing amber warning lamps (available from motorist's shops) some distance from your vehicle to give advance warning to other motorists. If your fault is not electrical and your vehicle has hazard warning flashing indicators, these should obviously also be put *on*.

If you are not technically minded or have no wish to carry out roadside running repairs then the best course of action is to join a motoring breakdown organisation.

Every car should contain a first aid kit and a fire extinguisher and drivers who are able to tackle their own repairs should carry — apart from the usual jack, handle and wheelbrace — an assortment of spanners, a torch, a flat piece of wood for placing under the jack (when forced to jack-up on soft ground), spare belts — fan or generator, a hose or two, jump leads, a tow rope, sparking plugs, folding triangles, cones etc, and a suit of overalls.

There may also be other items you could carry to suit your own particular vehicle.

ENGINE STALLS

A common cause of petrol-driven cars breaking down is an excessively rich mixture. This causes the engine to stall and very often occurs in cold weather when full-choke is required to start the engine. Should it be a manually operated choke the driver may have forgotten to ease in the choke progressively, or it can be caused by a defect in an automatically operated choke. If the vehicle has to stop for some while, such as at busy traffic lights, the petrol mixture can get richer and richer until it is too rich for combustion and the engine stalls (as it gets richer the mixture excludes

Below left:
All cars should carry a first aid kit and a fire extinguisher.

Below:
It is always a good idea to carry some basic tools and spares which could get you out of trouble if a breakdown occurs.

almost all the oxygen and the vapour of petrol only is a non-combustible gas).

Symptoms are that the car is all right while being driven along the road but, while halted at traffic-lights with the engine idling, the engine suddenly stops for no apparent reason.

To remedy the problem the gas has to be cleared from the cylinders so, with the hand-brake applied and the gear-lever in neutral, depress the accelerator pedal to the floor or to its maximum opening, then operate the starter for, say, ten seconds. You may need to do this two or three times, keeping the throttle depressed. The engine should then burst into life at very high revs. Progressively reduce the throttle opening and lower the engine revolutions.

If the car is fitted with an automatic choke and the trouble persists, you may complete your journey by keeping the revs high ie. maintain a very fast idle speed whenever you are required to halt, but do get the fault rectified as soon as possible.

Carburetter Flooding

This fault is very similar to the previous one and the symptoms are the same. It occurs if the carburetter float has punctured or there is dirt under the needle valve. Follow the aforementioned drill and keep the engine at a fast idle when stationary until you can get the fault rectified.

Above:
Occasionally disconnect and reconnect terminal blocks such as this to reduce the chance of an electrical failure through contacts becoming corroded.

Right:
Although it may not look it, this rear tyre is 10lbs below the recommended pressure and could lead to a blow-out with dire consequences.

Far right:
Driving over kerbs should be avoided as it can cause permanent damage and deflection in a tyre.

Left:

The car has a puncture in its off-side rear tyre but it would be dangerous to attempt to change the wheel here because of oncoming traffic.

ELECTRICAL PROBLEMS

An analysis of breakdown faults has shown that over 25% are caused by electrical failures. Although reasons can be varied, especially on the modern car, a recurring fault is corrosion forming at a terminal block, or at a spade connection which has not been disconnected for a long time.

It is a good idea when inspecting the engine compartment and elsewhere that connections occur (such as terminal blocks) to occasionally loosen or to disconnect and connect them several times to ensure that there is little likelihood of corrosion causing a bad connection at that point. Complete engine failure can occur as a result of corroded connections and what may seem like a major and costly fault simply turns out to be this.

BURST TYRES

When a tyre bursts the main danger is that the driver is shocked and the burst tyre causes the car to swerve and weave. It is important not to brake as this can cause skidding, and certainly avoid heavy braking.

Hold the steering wheel very firmly to check any swerving – but beware of over-correction. The aim is to let the car lose speed and roll to a safe stopping place.

The most likely reason for having a blow-out, which usually occurs at speed, is that the tyre is under-inflated.

The next most common cause is that the tyre has struck an object at sometime or has been driven over a kerb, perhaps for parking purposes.

Ford

CHANGING A TYRE

If a puncture should occur on the off-side (the driver's side on right-hand drive cars) and you are in a busy town thoroughfare endeavour to get to a quieter side street and then drive to the other side of the road so that the punctured wheel is by the kerb. This is to protect you from oncoming traffic; many motorists have been killed while changing a wheel on the side exposed to passing traffic. Whatever side of the car sustains a puncture make sure that this side is in the kerb before changing the wheel. Changing a wheel is straightforward enough and is explained in the driver's handbook, but if on soft ground remember to place your flat piece of wood beneath the jack base.

BEING TOWED

If your car develops a fault which necessitates being towed there are different ways to go about it. In the case of an engine failure there will be no power assistance with the footbrake and you will

Far left:

The driver has driven to the other side of the road to protect herself from oncoming traffic whilst she changes the off-side wheel. Note she is using her flat piece of wood beneath the jack because of the soft grass verge.

Left:

Place a card on the rear window of the defective car to show that it is *on tow* and quote the towing vehicle's registration number.

Left:

Fix the rope to the rear of the towing vehicle and to the front of the towed vehicle. Modern tow-ropes do not require knots.

Right:
The ignition key must be switched *on* to unlock the steering and allow other systems to function throughout the tow.

need to be very careful; in the case of an electrical fault you may have no headlamps or direction indicators.

Let's assume it is daylight, that the fault is engine failure and that we are going to use a tow rope — although a solid tube or rod gives a better tow.

Place a notice on the rear car's window warning other motorists that you are on tow and showing the towing car's registration number.

Connect the tow rope to both vehicles' towing eyes.

The towing driver must make clear to the towed driver his intentions and do so early so that the towed driver can in turn inform the following traffic of the intended manoeuvre.

Make sure that the towed vehicle's ignition is turned *on*. This releases the steering lock and enables the direction indicators, stop lights, headlamps and warning lights to be operational, if required.

Below right:
Keep the rope taut during towing and when stationary to prevent 'snatching' and the likelihood of snapping the rope as you move off.

Left:

The towing vehicle's driver should give clear signals, including hand signals, in plenty of time to complete a right-hand manoeuvre.

Remember that with engine failure you will have no braking assistance from the engine. This means great care needs to be exercised to prevent the towed vehicle colliding with the towing vehicle. The only braking effort will be that applied physically to the brake pedal.

The art of a successful tow is for the driver of the defective vehicle to stop his vehicle catching up with the towing vehicle. He must keep the tow rope taut at all times to stop snatching which may snap the tow rope. When stopping, you must keep the tow rope taut so that the towing vehicle pulls away smoothly without a snatch.

The driver of the towing vehicle must bear in mind that when making any change in direction or undertaking any other manoeuvre, sufficient time must be allowed for the two vehicles to complete the manoeuvre.

AUTO REPAIRS
OXFORD MEWS. BEXLEY HIGH St.
BREAKDOWN
24 HR.
NIGHT 0860 393473
AUTO REPAIRS
0322 54721

If these points are borne in mind a successful towing operation should be completed.

Note 1: if the breakdown is such that the engine can be run, then you must run the engine throughout the tow to give yourself brake assistance.

Note 2: some cars with automatic transmission cannot be towed. They will need to be recovered by a breakdown vehicle with a hoist tow.

JUMP LEADS

During the winter months one of the commonest problems experienced by motorists is a flat battery.This is usually due to the continuous use of headlights, fan heater, demisters and other electrical components coupled with little mileage.

Assistance from a fellow motorist and the use of jump-leads provides one solution. The following procedure should be followed:

Note: batteries give off hydrogen so for safety's sake a connection is made away from the flat battery. (5a/b refers)

1. See that any jewellery from the wrists or hands is removed before commencing.
2. Keep the two cars as far apart as possible from each other.
3. Check that both batteries are of the same voltage and polarity. If one car is negative earth and the other is positive – go no further.
4. Check the flat battery and top up with distilled water if necessary.

5(a) For both cars – POSITIVE EARTH
Taking one lead at a time, connect one end of the BLACK lead from the NEGATIVE terminal of the supply battery to the NEGATIVE terminal of the flat battery without allowing the lead to touch any metal part of either car. Then connect the RED lead from the POSITIVE terminal of the supply battery to a bright metal component or a good earthing point about 0.5 metre from the flat battery so that any sparking which may take place when making the final connection will be away from both batteries.

5(b) For both cars – NEGATIVE EARTH
Taking one lead at a time, connect one end of the RED lead to the POSITIVE terminal of the supply battery and the other end to the POSITIVE terminal of the flat battery without allowing the lead to contact any metal part of either car. Now connect one end of the BLACK lead to the NEGATIVE terminal of the supply battery and the other end of the same lead to a bright metal component or a good earthing point at least 0.5 metre from the flat battery. (This

Far left:
Because this car is fitted with an automatic transmission it had to be recovered with a hoist tow.

Above:
Ensure that maximum distance is maintained between the cars before making the final connection.

Left:
Make the final connection on the defective car away from the battery at a good, bright metal earthing point.

Right:
Run the serviceable car's engine at a fast idle whilst trying to start the defective car's engine.

ensures any sparking which may take place when making the final connection will be well away from both batteries)

6. Start the engine of the assisting car and leave running at a fast idle.

7. Now try to start the defective car's engine. If it doesn't start fairly quickly, do not persist as the problem is unlikely to be the battery.

8. Assuming the defective car's engine starts, allow it to warm-up. Now stop the other car's engine and remove the jump leads in the reverse order to which they were connected.

Warning:

The above method is the safest to adopt but it is common practice to connect the positive on the supply battery (slave battery) to the positive on the flat battery and then to connect the supply battery negative to the flat battery negative.

Take care if carried out in a confined area, such as in a garage where petrol vapour may be present, for if sparking is produced there could be trouble.

BREAKDOWN ON A MOTORWAY

If you are travelling on a motorway and your car develops a fault try to glide to a stop on the hard shoulder.

All motorways have emergency phones placed at intervals of about a mile. These are fixed on numbered posts and connected to the police authorities.

Numbered marker posts are placed at intervals of about 100 metres and each has an arrow indicating in which direction the nearest phone can be found.

You should use the phone to inform the police of your predicament.

Protecting Yourself

There has recently been concern over the number of fatal accidents that have befallen drivers waiting in or beside their stranded vehicles. As a result, the authorities now advise drivers to wait if possible, on an enbankment or on a grass verge well away from their vehicle and the passing traffic. Or, as some motoring organisations suggest, sit in the front passenger seat with the door wide open ready to vacate at any sign of danger.

Far left:

Marker posts are placed at intervals of 100m between phones. Each is numbered and indicates in which direction is the nearest phone.

Above:

A motorist has pulled on to the hard shoulder, fortuitously near an emergency phone.

Left:

These are the instructions you are asked to follow when using the emergency phone.

Authorities are also advising lady motorists to lock themselves in their cars and wait for the police to come along. This latest thinking is because of recently reported assaults on ladies whilst waiting beside their cars for help to arrive.

But it really is for the individual to decide what best to do under the circumstances and to be alert to any danger to themselves.

Lady motorists, however, may feel safer locked in their vehicle.

Minor Repairs

If you merely have a puncture then you are permitted to fit your spare wheel while on the hard shoulder providing you are not causing an obstruction to other road users. In fact you may carry out any minor repairs — such as changing a sparking plug – which can be executed in less time than it would take for a rescue vehicle to reach you.

Right:

This driver has stopped at an angle to give himself some protection from other traffic while he investigates the problem.

10 The MoT Test

INTRODUCTION

At present, when a vehicle reaches three years of age (dating from its first registration date) it is required to undergo a Department of Transport (formerly the Ministry of Transport) vehicle inspection for road worthiness. If it passes the inspection, a Vehicle Test Certificate, MOT certificate is issued which is valid for 12 months.

A car which has been used before its first registration ie: used abroad, must be tested three years from its date of manufacture. In practice the date of manufacture is taken to be the last day of the year in which the vehicle was built.

The MOT Certificate

An MOT certificate is not a guarantee that a vehicle is necessarily in a sound and good condition. All it amounts to is confirmation that at the time of the inspection all the items that have to be checked by law were found to be in a serviceable condition.

This certificate is important, however because it is illegal to use a car without one and you will be unable to obtain a renewal of your motor taxation licence (tax disc) if you do not possess a current MOT certificate for a car aged three years or more.

Remember that when applying for a motor taxation licence (see Chapter 4) you must produce a current insurance certificate and a current MOT certificate.

EXHAUST EMISSION TESTING

An important new element has recently been introduced to the MOT test. It consists of an analysis of the exhaust gases and although an engine may appear to be working perfectly; when the exhaust gases are analysed the car may prove to be a test failure.

This means that whereas in the past a vehicle may have passed its test and been presumed serviceable on all other accounts, this may not now be the case.

The object of this part of the test is to improve the quality of the air we breathe by reducing harmful gas emissions.

Keep this Certificate safely — See notes overleaf

The Department of Transport

Test Certificate

Serial number **NM** 0214639

The motor vehicle of which the Registration Mark is: B289 WNO

having been examined under section 45 of the Road Traffic Act 1988, it is hereby certified that at the date of the examination thereof the statutory requirements prescribed by Regulations made under the said section 45 were complied with in relation to the vehicle.

Vehicle Testing Station Number: 17279

Date of issue: OCTOBER 5TH 1991

Date of expiry: OCTOBER 7TH 1992

Serial Number of immediately preceding Test Certificate: MSO 178274

(To be entered when the above date of expiry is more than 12 months after the above date of issue.)

Approximate year of manufacture: 1984

Recorded mileage: 65999

If a goods vehicle, design gross weight: N/A kg

If not a goods vehicle, horse power or cylinder capacity of engine in cubic centimetres: 1600cc

Vehicle identification or chassis number: EC33135

Colour: RED

Make: FORD ESCORT

Signature of tester/inspector

Name in BLOCK CAPITALS: M WEBB.

WARNING
A Test Certificate should not be accepted as evidence of the satisfactory mechanical condition of a used vehicle offered for sale.

Authentication Stamp

CHECK
carefully that the particulars quoted above are correct. Certificates showing alterations should not be issued or accepted. They may delay the renewal of a Licence.

8251492.115682.7/91 VT20

Left:

You receive an MOT certificate such as this when your car has successfully passed its inspection.

Right:

A sign showing three white triangles on a blue background indicates that an establishment is an officially appointed MOT vehicle testing station.

First Step

At present only two of the exhaust gases are being checked – carbon monoxide (CO) and hydrocarbons (HC) — but this is sufficient to show whether engines are correctly tuned and running properly. Pass figures are less rigid for vehicles manufactured before 1983. In time as experience is gained and equipment is more highly developed we can expect more stringent testing. Cars already fitted with catalytic converters will also be tested and although normally should pass without any trouble, the test will detect a car with a defective converter. Diesel engined cars are not being analysed at present, partly because they produce fewer pollutants than even catalysed petrol cars but they do have a tendency to emit black smoke particles. These cars will have an instrumental smoke check to see if they are producing more diesel smoke than they need to, and partly because they require different test equipment and as the number of diesel cars is relatively small this cannot be justified.

Cars fitted with a catalytic converter may well fail an MOT test unexpectedly as defunct catalysts can be detected.

As a result of this additional check more failures and re-tests are envisaged.

Here are some simple checks that you should do before submitting your car for its Test:

Left:

Check that the direction indicators, horn and lights are all functioning correctly. This car has a faulty rear lamp which must be rectified immediately.

Below left:

These wiper blades are in a poor condition. They are not wiping clearly and the rubbers are split — a definite MOT failure.

Below right:

Top-up the windscreen-washer reservoir before checking that the jets are functioning properly.

Right:
Make sure that all the tyre pressures are correct before submitting the car for test.

ARRANGING A VEHICLE TEST

Officially appointed garages carry out the MOT inspection and very often these can be done while you wait. If your local garage does not operate this system the car should be booked in advance.

Preparing Your Car

Bear in mind that a car can fail an MOT test for a very minor defect, such as an empty windscreen-washer reservoir or a broken light bulb. It is therefore advisable to carry out a basic inspection beforehand

Check:

1: That all lights work and that headlamp beams are correctly angled.
2: That all wipers function correctly and that the blades are sound (certainly with no splits).
3: That seats are secure and that the belts are sound and not showing signs of fraying.
4: That doors and locks operate correctly.
5: That windscreens front and rear have no defects and give an absolutely clear view.
6: That Registration plates are secure and that lettering and numbers are legible and not obscured by tow-bars etc.
7: That windscreen washer jets are working properly.
8: That the body is sound with no sharp or jagged edges which might cause injury to other road users.
9: That you have the correct number of mirrors and that they are in good condition.
10: That the hazard warning light system is functioning correctly and that the hooter sounds audibly.
11: That there are no leaks from the fuel system and that the tank's fuel cap fits securely.
12: That tyres have at least 1.6mm depth of tread in a continuous band all the way round the circumference and across the full width of the tread and be inflated to their correct pressures.

These are very simple checks, the main point to remember is that you should not let your vehicle fail for some minor reason or other, that you should and could have done something about beforehand.

Pre-MOT Checks

If you are not technically minded there are garages who offer Pre-MOT checks and it may be worth investing in one of these. Don't leave it to the last minute, however, in case any repairs need to be carried out.

TEST FAILURE — AN APPEAL

If your vehicle has failed an MOT test you may appeal on form VT17, obtainable from any Testing Station or from a Vehicle Registration Office (VRO).

The completed form, with the current fee, must be received at the local VRO within 14 working days from 'refusal to issue a test certificate' date.

Left:

Many organisations will do a pre—MOT check and here an AA inspector is carrying out a pre-test examination.

DO NOT WRITE HERE

Notice of Appeal

Road Traffic Act 1972 Section 43

PLEASE USE BLOCK CAPITALS
BEFORE COMPLETING THIS NOTICE OF APPEAL PLEASE READ THE NOTES OVERLEAF.

*I/We, being aggrieved by the refusal of a test certificate, or the grounds upon which the certificate was refused, in respect of the motor vehicle particulars of which are given below, hereby appeal to the Secretary of State for Transport.

1. The grounds upon which the appeal is made are:

2. Particulars of vehicle

Registration No .... Make .... Type ....

If vehicle is a motor cycle, state whether solo or with sidecar & cubic capacity ....

Date of Manufacture ....

3. Name and full address of appellant

*Mr/Mrs/Miss or other title ....

....

Postcode .... Telephone number during normal working hours ....

4. Name, full address and number of Vehicle Testing Station responsible for issuing the refusal notice.

....

....

Vehicle Testing Station number ....

(The vehicle testing station number is given on the Notification of Refusal of a Test Certificate – see note 3 overleaf)

Indicate here those days of the week (excluding Saturday & Sunday) on which it will be convenient for the vehicle to be submitted for the appeal test – *see note 5 overleaf.*

Monday	*morning/afternoon	Thursday	*morning/afternoon
Tuesday	*morning/afternoon	Friday	*morning/afternoon
Wednesday	*morning/afternoon		

6. a. The Notification of Refusal of a Test Certificate No .... was issued on .... 19......
I have read note 6 overleaf and I declare that the change(s) which have taken place in the condition of the vehicle since that date are as follows
(please tick appropriate box)

No change(s) ☐ Some Change(s) ☐

b. If you have ticked the second box, describe the change(s):

7. A crossed *cheque/money order/postal order for £ .... is enclosed (see note 2 overleaf)

Signature of appellant or person acting for appellant .... Date ....

**Delete as appropriate*

VT17 (Rev 12/81)

You must not have your vehicle repaired or worked on in any way before an appeal test is carried out. Any change in the vehicle may affect the outcome.

If your appeal is successful the fee, or if appropriate part of it, may be refunded.

Payment Of Fees For Re-Test

The full fee is payable:
a. For a re-test at a station not employed to do the repair.
b. If a vehicle is submitted for a re-test more than 14 days after being failed.

Half-fee Is Payable:
a. If the vehicle is returned to the original testing station which carried out the test for repair and re-test within 14 days of being failed.

No Further Fee Is Payable:
a. If the vehicle is left for repair and re-test at the testing station which carried out the test.

AN OFFENCE

It is an offence to use your vehicle on the highway if it is of 'testable age' and does not have a current test certificate.

Exceptions are:
a. Taking it away from a testing station after it has failed.
b. Taking it to or away from a place whereby previous arrangements have been made to repair the defect for which the vehicle was failed.
c. Taking it to a testing station for its test already booked in advance.

Even in the above circumstances it is possible that you may be prosecuted for driving an unroadworthy vehicle if it does not comply with the various regulations regarding its construction and use.

Also, remember that your insurance may not be operative under these conditions.

Finally, if you do find that you have let your MOT certificate lapse and it is out of date, you may use your car on the road, 'legally' if you are driving to a Test Station for a *pre-arranged test.*

Left:
If your car does fail its test and you wish to appeal, make your appeal on this form (VT17).